KU-738-027

Measurements and their Uncertainties

A practical guide to modern error analysis

IFAN G. HUGHES

Physics Department, Durham University

THOMAS P. A. HASE

Physics Department, Warwick University

GEORGE GREEN LIBRARY OF
SCIENCE AND ENGINEERING

OXFORD

UNIVERSITY PRESS

OXFORD
UNIVERSITY PRESS

Great Clarendon Street, Oxford OX2 6DP

Oxford University Press is a department of the University of Oxford.
It furthers the University's objective of excellence in research, scholarship,
and education by publishing worldwide in

Oxford New York

Auckland Cape Town Dar es Salaam Hong Kong Karachi
Kuala Lumpur Madrid Melbourne Mexico City Nairobi
New Delhi Shanghai Taipei Toronto

With offices in

Argentina Austria Brazil Chile Czech Republic France Greece
Guatemala Hungary Italy Japan Poland Portugal Singapore
South Korea Switzerland Thailand Turkey Ukraine Vietnam

Oxford is a registered trade mark of Oxford University Press
in the UK and in certain other countries

Published in the United States
by Oxford University Press Inc., New York

© I. Hughes & T. Hase 2010

The moral rights of the authors has been asserted
Database right Oxford University Press (maker)

First published 2010

All rights reserved. No part of this publication may be reproduced,
stored in a retrieval system, or transmitted, in any form or by any means,
without the prior permission in writing of Oxford University Press,
or as expressly permitted by law, or under terms agreed with the appropriate
reprographics rights organization. Enquiries concerning reproduction
outside the scope of the above should be sent to the Rights Department,
Oxford University Press, at the address above

You must not circulate this book in any other binding or cover
and you must impose the same condition on any acquirer

British Library Cataloguing in Publication Data

Data available

Library of Congress Cataloging in Publication Data

Data available

Typeset by SPI Publisher Services, Pondicherry, India
Printed in Great Britain
on acid-free paper by
Clays Ltd, St Ives plc   1 00779 0657

ISBN 978–0–19–956632–7 (hbk)
       978–0–19–956633–4 (pbk)

10

10 0779065 7

MEASUREMENTS AND THEIR UNCERTAINTIES

UNIVERSITY OF NOTTINGHAM

WITHDRAWN

FROM THE LIBRARY

09

Er cof am Mam. (IGH)
For my parents. (TPAH)

Preface

This book grew out of the process of revamping the first-year practical course at the physics department at Durham University. During the restructuring of the laboratory course, we noted the significant changes in the skill set of the student cohort over the previous decade. We searched for a book that could be a recommended text for the treatment of uncertainties. There is no shortage of books which deal with uncertainties in measurements and error analysis. Most of these books treat error analysis in the traditional, old-fashioned approach which does not take into account modern developments—indeed, error propagation is often treated as an exercise in calculus of many variables. In modern laboratories computers are used extensively for data taking and analysis, and students now have access to, and familiarity with, spreadsheets for manipulation of data. This book is written assuming that most of the number crunching will be done by computer. Traditional textbooks have appendixes which list, e.g. Gaussian integrals, cumulative distribution functions, tables of chi-squared distribution likelihoods, and so on. Our emphasis is on calculating the relevant numbers within routinely available spreadsheet software packages. In contrast to traditional books, we have decided to produce a hands-on guide book: key points illustrated with worked examples, concise and in a handy format— sufficiently user-friendly that students actually bring the book along and use it in the teaching laboratory.

The scope of this book is to cover all the necessary groundwork for laboratory sessions in a first- and second-year undergraduate physics laboratory and to contain enough material to be useful for final-year projects, graduate students and practising professional scientists and engineers.

In contrast to the mathematically rigorous treatment of other textbooks we have adopted a 'rule of thumb' approach, and encourage students to use computers to assist with as many of the calculations as possible. Last century if a student had knowledge of an angle and its uncertainty, and was testing the validity of Rutherford's scattering law with its $\sin^{-4}(\theta/2)$ dependence, the suggested approach to error propagation was to turn this into an exercise in differentiation. Nowadays, a student can create a spreadsheet, calculate the value of the function at the desired angle, at the angle+error bar, and deduce the uncertainty in propagation through Rutherford's formula much more quickly. Throughout the book we encourage a functional approach to calculations, in preference to the calculus-based approximation. A large number of end-of-chapter exercises are included, as error analysis is a participation, rather than a spectator, sport.

The only prerequisite is suitably advanced mathematics (an A-level, or equivalent) which is a compulsory qualification for studying physics and engineering at most universities. Although written originally for use in a physics laboratory we believe the book would be a useful hands-on guide for undergraduates studying the physical sciences and engineering.

A preliminary version of this book ('the little red book') was used by more than 1500 students following the *Discovery Skills* module at Durham University; we are grateful to all of those who helped identify and eradicate typographical errors, inconsistencies and sources of confusion.

Several colleagues involved in the teaching laboratories at Durham have contributed to this book via a number of discussions. We would like to record our gratitude to Del Atkinson (we realise the book is 'full of errors'), Richard Bower, David Carty, Paula Chadwick, Simon Cornish, Graham Cross, Nigel Dipper, Ken Durose, David Flower, Douglas Halliday, Michael Hunt, Gordon Love, John Lucey, Lowry McComb, Simon Morris, Robert Potvliege, Steve Rayner, Marek Szablewski, and Kevin Weatherill. Many colleagues kindly donated their time to proofread various chapters, and we are indebted to them for this service, including Charles Adams, Matthew Brewer, Paula Chadwick, Stewart Clark, Malcolm Cooper, David Flower, Patrick Hase, Jony Hudson, Martin Lees, Phillip Petcher, Jon Pritchard, and Marek Szablewski. Simon Gardiner gave invaluable TEX advice. For any remaining flaws or lack of clarity the authors alone are responsible. While performing error analysis in research, the authors have benefited from discussions with Ed Hinds, Jony Hudson, Clemens Kaminski, Ben Sauer, Peter Smith, Derek Stacey, and Duncan Walker. Damian Hampshire pointed out that 'truth' should be sought in the department of theology, not an error-analysis book. Simon Cornish and Lara Bogart kindly provided us with data for some of the figures. We are grateful to Bethan Mair and Tony Shaw for advice on writing a book, Mike Curry for providing food and lodging, and Stewart Clark and Marek Szablewski for eradicating the loneliness of long-distance runs.

We would like to thank our families for support and encouragement, and Sönke Adlung and April Warman at OUP for their enthusiasm and patience.

December 2009   Durham and Warwick

Contents

Errors in the physical sciences

What is the role of experiments in the physical sciences? In his *Lectures on Physics*, Volume 1, page 1, the Nobel Laureate Richard Feynman states (Feynman 1963):

> The principle of science, the definition, almost, is the following: *The test of all knowledge is experiment*. Experiment is the *sole judge* of scientific 'truth'.

We will emphasise in this book that an experiment is not complete until an analysis of the uncertainty in the final result to be reported has been conducted. Important questions such as:

- do the results agree with theory?
- are the results reproducible?
- has a new phenomenon or effect been observed?

can only be answered after appropriate error analysis.

1.1 The importance of error analysis

The aim of error analysis is to quantify and record the errors associated with the inevitable spread in a set of measurements, and to identify how we may improve the experiment. In the physical sciences experiments are often performed in order to determine the value of a quantity. However, there will always be an error associated with that value due to experimental uncertainties. The sources of these uncertainties are discussed later in this chapter. We can never be certain what the exact value is, but the errors give us a characteristic range in which we believe the correct value lies with a specified likelihood.

These ideas do not just apply to the undergraduate laboratory, but across the entire physical sciences at the very fundamental level. Some of the so-called fundamental constants, in addition to other physical constants, have been determined experimentally and therefore have errors associated with them. For fundamental constants the best accepted values can be found from experiments performed at Government laboratories such as NIST (National Institute of Standards and Technology)[1] in the USA and the National Physical Laboratory (NPL)[2] in the UK.

As an example, the currently accepted values (the first number), and their errors (the number after the $\pm$), of Avogadro's constant, N_A, and the Rydberg constant, R_∞, are:

[1] http://www.nist.gov/index.html
[2] http://www.npl.co.uk

$$N_A = (6.022\,141\,79 \pm 0.000\,000\,30) \times 10^{23}\,\text{mol}^{-1},$$

and

$$R_\infty = (10\,973\,731.568\,527 \pm 0.000\,073)\,\text{m}^{-1}.$$

[3] An alternative way of writing the error is: $R_\infty = 10\,973\,731.568\,527\,(73)\,\text{m}^{-1}$, although we will exclusively use the $\pm$ convention in this book.

Note that although the Rydberg constant is known to an impressive precision (one part in 10^{11}) there is still an uncertainty in its value.[3] The only constants which do not have an error are those that have been defined. One of the most common is the speed of light in a vacuum, $c = 299\,792\,458\,\text{m s}^{-1}$ (exact). A full list of the physical constants and their errors is reviewed periodically by the International Council for Science: Committee on Data for Science and Technology (CODATA) task group on fundamental constants. A self-consistent set of internationally recommended values of the basic constants and conversion factors is derived from all the relevant data available, and is periodically updated. The latest review is available on the fundamental constants web-page hosted at NIST.[4]

[4] http://physics.nist.gov/cuu/Constants/

1.2   Uncertainties in measurement

How should we interpret an uncertainty in a measurement? Reporting a quantity as the best estimate $\pm$ the error should be regarded as a statement of probability. The scientists who performed the measurements and analysis are confident that the Avogadro constant is within the range $6.022\,141\,49 \times 10^{23}\,\text{mol}^{-1} \le N_A \le 6.022\,142\,09 \times 10^{23}\,\text{mol}^{-1}$. They cannot be certain that the Avogadro constant is within the limits quoted, but the measurements lead them to believe that there is a certain probability of its being so. In later chapters in this book we will use statistical analysis to define this range, and thereby its confidence level, more quantitatively.

Why do we need a statistical description of experiments performed on systems which are usually described by a well-known set of equations, and are hence deterministic? Figure 1.1 shows the results of an experiment measuring the range of a ball-bearing launched from a spring-loaded projectile launcher. Successive measurements were taken of the distance the ball-bearing projectile landed with respect to the launching cannon. A histogram of the projectile distance is plotted in Fig. 1.1(b). The crucial point to note is the following: although nominally these repeat measurements are performed under exactly *the same conditions*, and Newton's laws of motion, which govern the trajectory, are time independent, *successive repeats of the experiments gave different values* for the range of the ball-bearing projectile.

The histogram in Fig. 1.1(b) contains the information about the range of the projectile, and the uncertainty in this range. In this book we will discuss statistical techniques to analyse the location of the centre and width of histograms generated from multiple repeats of the same experiment. For this example these techniques enable a quantitative determination of the range and its uncertainty.

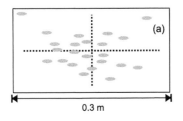

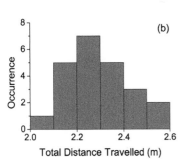

Fig. 1.1 The range of a ball-bearing launched from a spring-loaded projectile launcher. (a) shows the location at which the projectile landed on carbon paper—the repetition of the experiment with nominally the same experimental conditions is seen to yield different results. In (b) a histogram of the total radial range of the projectile from the launcher is constructed, with the number of occurrences within a bin of width 10 cm plotted.

1.2.1   Terminology

You should note that despite many attempts to standardise the notation, the words 'error' and 'uncertainty' are often used interchangeably in this context—this is not ideal—but you have to get used to it! (A discussion of the International Standardisation Organisation's *Guide to the Expression of Uncertainty in Measurement (GUM)* is presented in Chapter 9.)

There are two terms that have very different meanings when analysing experimental data. We need to distinguish between an **accurate** and a **precise** measurement. A precise measurement is one where the spread of results is 'small', either relative to the average result or in absolute magnitude. An accurate measurement is one in which the results of the experiment are in agreement with the 'accepted' value. Note that the concept of accuracy is only valid in experiments where comparison with a known value (from previous measurements, or a theoretical calculation) is the goal—measuring the speed of light, for example.

Figure 1.2 shows simulations of 100 measurements of a variable. The dashed vertical line in the histogram shows the accepted value. The scatter of the data is encapsulated in the width of the histogram. Figures 1.2(a) and (c) show examples of precise measurements as the histogram is relatively narrow. In Fig. 1.2(a) the centre of the histogram is close to the dashed line, hence we call this an accurate measurement. The histograms in Fig. 1.2 show the four possible combinations of precise and accurate measurements: Fig. 1.2(a) represents an accurate and precise data set; Fig. 1.2(b) an accurate and imprecise data set; Fig. 1.2(c) an inaccurate but precise data set; and finally Fig. 1.2(d) both an inaccurate and imprecise data set.

Based on the discussion of precision and accuracy, we can produce the following taxonomy of errors, each of which is discussed in detail below:

- **random errors**—these influence precision;
- **systematic errors**—these influence the accuracy of a result;
- **mistakes**—bad data points.

1.2.2   Random errors

Much of experimental physics is concerned with reducing random errors. The signature of random errors in an experiment is that repeated measurements are scattered over a range, seen in Fig. 1.1. The smaller the random uncertainty, the smaller the scattered range of the data, and hence the more precise the measurement becomes. The best estimate of the measured quantity is the **mean** of the distributed data, and as we have indicated, the error is associated with the distribution of values around this mean. The distribution that describes the spread of the data is defined by a statistical term known as the **standard deviation**. We will describe these terms in greater detail in Chapters 2 and 3.

Having quantified the uncertainty in a measurement, the good experimentalist will also ask about the origin of the scatter of data. There are two categories of scatter in experiments: (1) technical, and (2) fundamental noise.

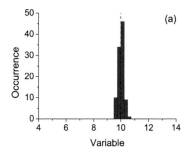

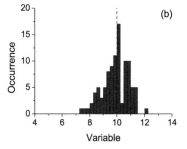

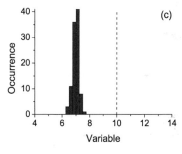

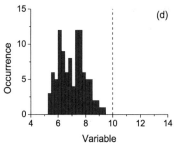

Fig. 1.2 Terminology used in error analysis. Simulations of 100 measurements are shown in histograms of constant bin-width. The extent of the scatter of the data (the width of the histogram) is a measure of the precision, and the position of the centre of the histogram relative to the dashed line represents the accuracy. The histograms show (a) precise and accurate, (b) imprecise and accurate, (c) precise and inaccurate and (d) imprecise and inaccurate sets of measurements.

EwMDA1QFAGpr

A given apparatus will have a fundamental noise limit (governed by some law of physics, e.g. diffraction) but will typically operate with a higher noise level (technical noise) which could, in principle, be reduced. In Fig. 1.1 the source of the scatter is technical in nature. The trajectories are subject to the deterministic Newton's laws of motion. However, the parameters which categorise the trajectories can vary. Maybe the launcher recoils and changes the next launch angle, or perhaps reloading the cannon with the same projectile might load the spring to a different tension each time. It is possible to design experiments with sufficient care that the origin of the random errors approaches the fundamental noise limit of the apparatus. The good experimentalist will take measures to reduce the technical noise for a given apparatus; if the fundamental noise level remains the limiting feature of the experiment a new apparatus or approach is needed.

Examples of fundamental noise limits include Johnson and shot noise. Johnson noise[5] arises in a resistor as a consequence of the thermal agitation of the charge carriers and becomes important in experiments measuring low voltages. The discrete nature of certain quantities gives rise to so-called shot noise—when measuring feeble light beams the manifestation of the electromagnetic field being composed of finite energy bundles, or photons, gives rise to random fluctuation of the intensity.[6]

There exist many statistical techniques for quantifying random errors, and minimising their deleterious effect. These will be discussed at length in later chapters. It should be emphasised that these techniques all rely on taking repeated measurements.

1.2.3   Systematic errors

Systematic errors cause the measured quantity to be shifted away from the accepted, or predicted, value. Measurements where this shift is small (relative to the error) are described as accurate. For example, for the data from the projectile launcher shown in Fig. 1.1 the total range needs to be measured from the initial position of the ball-bearing within the launcher. Measuring from the end of the cannon produces a systematic error. Systematic errors are reduced by estimating their possible size by considering the apparatus being used and observational procedures.

In Figs. 1.2(c) and (d) the measurements are scattered, but none of the individual measurements is consistent with the accepted value: the measurements are consistently smaller than the predicted value. This is the tell-tale sign of at least one, and possibly more, systematic error(s). In contrast to random errors, there do not exist standard statistical techniques for quantifying systematic errors. It is left to the experimenter to devise other sets of measurements which might provide some insight as to the origin of the systematic discrepancies.

1.2.4   Mistakes

Another class of error which defies mathematical analysis is a mistake. These are similar in nature to systematic errors, and can be difficult to detect. Writing

[5]Named after J. B. Johnson, who first studied the effect at Bell Laboratories; *Thermal agitation of electricity in conductors*, Nature (1927) **119**, 50–51.

[6]Even for a constant-intensity light source, the actual number, N, of photons detected in a given time will have a Poisson distribution about the (well-defined) mean. We will show in Chapter 3 that the photon noise is equal to the square root of the average number of photons, $\sqrt{N}$. The signal-to-noise ratio is then $\dfrac{N}{\sqrt{N}} = \sqrt{N}$. When the number of photons collected is small the shot-noise limited signal-to-noise ratio can be the limiting feature of an experiment.

2.34 instead of 2.43 in a lab book is a mistake, and if not immediately corrected is very difficult to compensate for later.

There are other well-known types of mistakes which can influence the precision and accuracy of experimental results with potentially disastrous consequences. **Misreading scales** occurs often with an analogue device which has a 0–10 scale above the gradations, and a 0–3 scale underneath. Care has to be taken when using a signal generator where an analogue dial from 1 to 10 is used in conjunction with multiplier knobs such as 1–10 kHz. With an instrument such as an oscilloscope one has to be careful to check whether a **multiplier** such as $\times 10$ has been engaged. In Fig. 1.3 a histogram of events is shown, where the automated data-collecting software misfired on 10% of the events. **Malfunction of the apparatus** can be difficult to spot; the presence of erroneous points can become apparent when the data are displayed graphically. There are many examples where **confusion over units** has had disastrous consequences. In 1999 the failure of the NASA Mars Climate Orbiter was attributed to confusion about the value of forces: some computer codes used SI units, whereas others used imperial. A Boeing 767 aircraft ran out of fuel mid-flight in 1983; a subsequent investigation indicated a misunderstanding between metric and imperial units of volume.

Obviously the good experimentalist makes very few, if any, such mistakes; their effects will not be discussed further in this book.

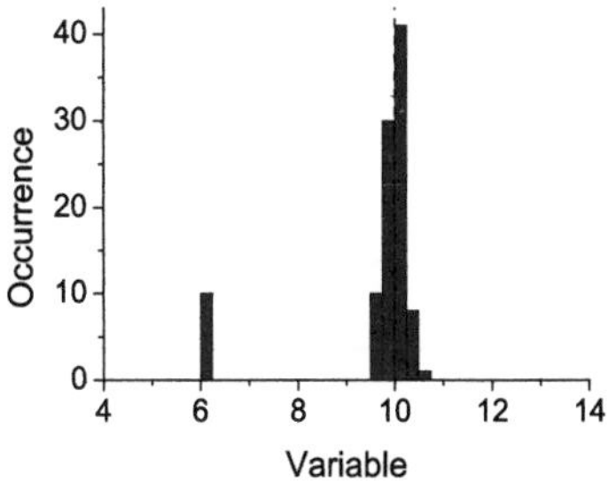

Fig. 1.3 A histogram of measurements when 10% of the events are logged incorrectly by automated data-collecting software. The presence of the erroneous points is clearly visible when the data are displayed graphically.

1.3   Precision of measurements

There are certain measurements with no statistical scatter; repeating the experiment does not yield more useful information. If six successive measurements of the amount of acid titrated are 25.0, 25.0, 25.0, 25.0, 25.0 and 25.0 cm^3 it is obviously a waste of time to perform another similar measurement. In this case the precision of the measurement is limited by the finite resolution of the scale on the titration apparatus. How do we estimate the precision of the device in this case? We discuss the two types of measuring instrument (analogue and digital) below.

It should be stressed that estimating the uncertainty based on some property of the measuring device is only valid if successive measurements are identical. The spread of the ranges in Fig. 1.1 is significantly greater than the precision of the measuring device, hence the statistical techniques introduced in the next chapter must be used to calculate the uncertainty in that case.

RULE OF THUMB: The precision of a measurement only equals the precision of the measuring device when all repeated measurements are identical.

1.3.1   Precision of an analogue device

Imagine that you are measuring the length of a piece of A4 paper with a ruler with 1 mm gradations. Successive measurements all give 297 mm. It seems reasonable to estimate the precision in this case to be half a division, i.e. ± 0.5 mm, and thus we would report the length of the piece of paper as 297.0 ± 0.5 mm.

It is also worth considering cases where this rule of thumb is too pessimistic. Imagine that the measurements of the length of the A4 paper were performed

RULE OF THUMB: The highest precision achievable with an analogue device such as a ruler is half a division.

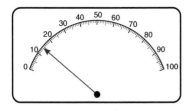

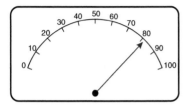

Fig. 1.4 The upper part of the figure displays a situation where estimating the uncertainty to be half a division is appropriate; in contrast in the lower part of the figure the uncertainty in the measurement is substantially smaller than half a division.

RULE OF THUMB: The precision of a digital meter is limited to the last decimal point; i.e. one in the last digit.

RULE OF THUMB: Ensure apparatus is properly calibrated and zeroed.

with a ruler which only had 1 cm gradations. All measurements of the length would lie between 29 and 30 cm; in this instance estimating the uncertainty to be half a division, or ± 0.5 cm, is a gross overestimate. A good experimentalist will be able to interpolate, i.e. estimate the position with a finer resolution than the gradations. There is no simple rule as to what value to report for the precision in this case; indeed, the error estimation can only be done by the experimenter, and is likely to vary for different experimenters. Figure 1.4 illustrates the issues associated with the precision of an analogue device.

1.3.2   Precision of a digital device

Repeat measurements of the voltage of a battery with a digital multimeter each yields 8.41 V. What is the uncertainty?

Some digital instruments come with manufacturer's specifications for the uncertainty, such as 'half the last digit'—the equivalent result to the analogue case. Therefore one could write 8.410 ± 0.005 V. Note the extra zero which appears at the end of the reported number which is not present in the actual reading. There is a significant assumption implicit in this estimate, namely that the digital instrument does an appropriate rounding, i.e. it does not truncate the number. If the former occurs then 8.419 would appear as 8.42, whereas if it is the latter, 8.419 would appear as 8.41. Ascertaining whether the meter rounds or truncates can be difficult, therefore the conservative estimate is to use the full last digit, i.e. we would report 8.41 ± 0.01 V for the example above to be on the safe side.

1.4   Accuracy of measurements

The accuracy of an experiment is determined by systematic errors. For an inaccurate set of measurements there will be a difference between the measured and accepted values, as in Figs. 1.2(c) and (d). What is the origin of this discrepancy? Answering this question, on the whole, is difficult and requires insight into the experimental apparatus and underlying theories. For the example of the projectile launcher, the range is a function of both the launch angle and muzzle speed. Experimental factors which could affect the accuracy include the setting of the launch angle or the reliance on the manufacturer's specification of the muzzle speed. The theoretical prediction, for this example, is based on a calculation which ignores air resistance—the validity of this assumption could be questioned.

Three of the more common sources of systematic error are **zero, calibration and insertion errors**. An example of a zero error is using a ruler to measure length if the end of the ruler has been worn away. A metal ruler calibrated at 20 °C will systematically yield measurements which are too large if used at 10 °C owing to the thermal contraction of the gradations; this is a calibration error. Examples of insertion errors include: placing a room-temperature thermometer in a hot fluid, which will change the temperature of the fluid; the current in an electrical circuit being changed by placing a meter across a component.

Chapter summary

- An experiment is not complete until an appropriate error analysis has been conducted.
- When successive measurements of the same quantity are repeated there is usually a distribution of values obtained.
- A crucial part of any experiment is to measure and quantify the uncertainties in measurement.
- An accurate measurement agrees with the expected value.
- A precise measurement has a small relative uncertainty.
- The signature of the presence of random errors is that repeat measurements of the same quantity produce different results.
- The statistical spread of a data set is a reflection of the precision of the measurement.
- The deviation of the centre of the histogram from the accepted value is a reflection of the accuracy of the measurement.
- Systematic errors influence the accuracy of a measurement.
- The precision of an analogue device is half a division.
- The precision of a digital device is one in the last digit.
- The precision of a measurement is only equal to the precision of the measuring device if repeat measurements are identical.
- Ensure apparatus is properly calibrated and zeroed.

Random errors in measurements

In Section 1.2 we discussed how random errors are a feature of experiments in the physical sciences. In the absence of systematic errors there is a spread of measurements about the accepted value owing to random errors. We have seen in Chapter 1 that when successive measurements of the same quantity are repeated there is a distribution of values. A reading taken at a given time will differ from one taken subsequently. In Section 1.2 we discussed some of the possible causes for these fluctuations. In this chapter we take it as given that there are random errors in repeated measurements, and discuss techniques of how to analyse a set of measurements, concentrating in particular on how to extract objective numbers for two vital properties of a distribution: (i) our best estimate of the quantity being measured, and (ii) our estimate of the uncertainty in the value to report. We employ statistical techniques for the analysis to reflect the fact that the distribution of measurements is a consequence of statistical fluctuations inherent in collecting a finite number of data points; we implicitly assume that there are no systematic errors.

2.1 Analysing distributions: some statistical ideas

In Fig. 2.1 we plot histograms of the occurrence of a particular value of a measured quantity, x. Four histograms are shown where the number of data points collected, N, increases from (a) 5, to (b) 50, to (c) 100, to (d) 1000. It is apparent that as the number of data points increases, the distribution becomes smoother, but that the width remains unchanged. A smoother histogram will facilitate a more precise estimate of the three quantities which are of most interest to us: the **centre**, the **width** and the **uncertainty in the location of the centre**. We discuss methods for ascertaining the value of each of these quantities in the following sections.

2.2 The mean

The best method for reducing the effects of random errors on a measurement is to repeat the measurement, and take an average. Consider N measurements, $x_1, x_2, \ldots, x_N$. The fluctuations responsible for the spread of readings are

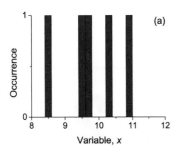

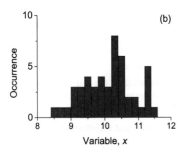

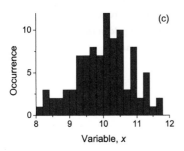

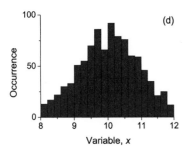

Fig. 2.1 Histograms of different experimental runs with (a) 5, (b) 50, (c) 100 and (d) 1000 data points. The histograms become smoother as more data points are collected, making it easier to deduce the centre, the width and the uncertainty in the location of the centre.

random, consequently they are equally likely to be higher as lower than the accepted value. The arithmetic mean is a way of dividing any random errors among all the readings. We therefore adopt the mean, $\bar{x}$, as our best estimate of the quantity x.

$$\bar{x} = \frac{1}{N}(x_1 + x_2 + \cdots + x_N) = \frac{1}{N}\sum_{i=1}^{N} x_i. \tag{2.1}$$

However, as is evident from Fig. 2.1, quoting the mean does not yield all the information that is inherent in a set of repeat measurements.

2.3   The width of the distribution: estimating the precision

The precision of the N measurements can be estimated from the distribution of the scattered data. The more similar the readings, the smaller the width of the distribution, and the more precise the measurement. (This was the situation depicted in Fig. 1.2a and b.) We can therefore quantify this precision by looking in detail at the width of this distribution.

2.3.1   Rough-and-ready approach to estimating the width

Suppose we had timed the period of oscillation, T, of a pendulum 10 times and obtained the values listed in Table 2.1.

Table 2.1 Ten measurements of the period of oscillation, T, of a pendulum. The precision of the measuring device is 0.1 s.

Trial	1	2	3	4	5	6	7	8	9	10
Period/s	10.0	9.4	9.8	9.6	10.5	9.8	10.3	10.2	10.4	9.3

All data points lie within the interval $9.3 \leq T \leq 10.5$ s which covers a range of 1.2 s, or a spread around the mean ($\bar{T} = 9.9$ s) of about ± 0.6 s. Evaluating the maximum spread of the data is one rough-and-ready approach to estimating the precision of the measurement. It is somewhat pessimistic, because as we will see later, we generally take the precision in a measurement to be two-thirds of the maximum spread of values, and therefore the spread of the data in Table 2.1 is approximately ± 0.4 s. The factor of two-thirds is justified for a Gaussian distribution, discussed in Chapter 3. Note that the spread of the measurements is significantly worse than the precision of the measuring instrument—this is why taking repeat measurements is important.

We can therefore say that:

- a typical measurement of the period is likely to be within 0.4 seconds of the mean value;
- the precision of the measurements is 0.4 seconds.

In many cases this simple procedure (illustrated in Fig. 2.2a) provides a fast method for estimating the precision, and is adequate in estimating the spread of the data around the mean. It is particularly appropriate when the number of data points is low, $N \leq 10$. The method is summarised in the box below:

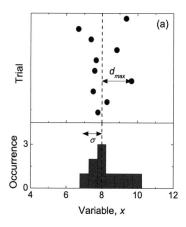

Rough-and-ready approach to estimating the width

- Calculate the mean of the data set, $\bar{x}$.
- Calculate the maximum spread of the data around the mean: either $d_{max} = x_{max} - \bar{x}$, or $d_{max} = \bar{x} - x_{min}$, whichever is greater.
- Quote the standard deviation as two-thirds of the maximum deviation: $\sigma = 2/3 \times d_{max}$.

2.3.2 Statistical approach to estimating the width

As more data points are recorded there is an increasing probability that the distribution describing the spread will have extended wings. Calculating the spread of data from the maximum deviation can give a misleading estimation (illustrated in Fig. 2.2b). To avoid this problem we need a measure of the random uncertainty that depends on *all the measurements*, and not just the extreme values as in the method above. The statistical quantity that is widely used is the **standard deviation**.

We define the deviation from the mean of the ith data point as:

$$d_i = x_i - \bar{x}. \tag{2.2}$$

The deviation d_i is equally likely to be positive as negative, and therefore when summed over a data set will be zero. Returning to the distributions of the period of oscillation from Table 2.1, Table 2.2 shows the 10 measurements and their deviation from the mean.

The average deviation, $\bar{d} = (1/N) \sum_i d_i$, is zero and therefore cannot be used as a measure of the spread of the data. Note that in Table 2.2 the average of the squared deviation is not zero. This motivates the introduction of the sample **variance**, which is the square of the standard deviation, σ, as the sum of the squares of these deviations over the data set. The standard deviation is

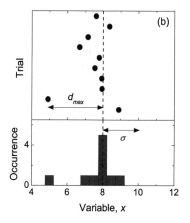

Fig. 2.2 Histograms of 10 experimental measurements of the variable x. In (a) the maximum deviation d_{max} is positive, and the width of the distribution, σ, is also indicated. For (b) the point furthest from the mean has a negative deviation, and the width is taken as two-thirds of the modulus of this deviation. In (b) the estimate of the precision is dominated by one deviant point.

Table 2.2 Ten measurements of the period of oscillation, T, of a pendulum. The sum of the deviations $\Sigma_i d_i = 0$, and the sum of the squares of the deviations $\Sigma_i d_i^2 = 1.58 \text{ s}^2$.

Period, T (s)	10.0	9.4	9.8	9.6	10.5
Deviation, d (s)	0.07	−0.53	−0.13	−0.33	0.57
Squared deviation, d^2 (s^2)	0.0049	0.2809	0.0169	0.1089	0.3249
Period, T (s)	9.8	10.3	10.2	10.4	9.3
Deviation, d (s)	−0.13	0.37	0.27	0.47	−0.63
Squared deviation, d^2 (s^2)	0.0169	0.1369	0.0729	0.2209	0.3969

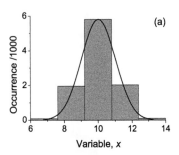

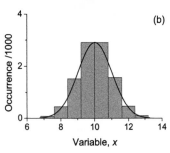

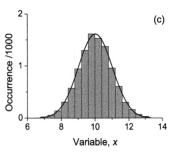

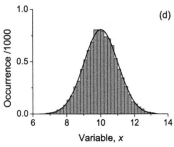

Fig. 2.3 Histograms of 10 000 samples with mean 10 and standard deviation 1. The bin-width is (a) 1.6, (b) 0.8, (c) 0.4 and (d) 0.2. The histograms become smoother as the bin-width decreases, tending toward a smooth distribution. Each histogram also has a normal-distribution curve superimposed. The agreement between the discrete and continuous functions improves as the bin-width is reduced.

thus defined:

$$\sigma_{N-1} = \sqrt{\frac{(d_1^2 + d_2^2 + \cdots + d_N^2)}{N-1}} = \sqrt{\frac{1}{N-1}\sum_{i=1}^{N} d_i^2}. \tag{2.3}$$

The distribution is normalised to $(N-1)$ as opposed to N (as was the case for the mean) because the data have already been used once to find the mean and there are thus only $(N-1)$ independent values of the deviations with which to define the variance (see Chapter 8).

The standard deviation is thus a statistical measure, based on *all of the available data*, of the width of the distribution and hence will be a more reliable estimation of the precision. When we take a series of measurements, we obtain a mean value, $\bar{x}$, as our best estimate from that set of measurements, with a standard deviation σ. If we were to make one more measurement, we believe there is a two-thirds chance that the new measurement will lie in the interval $\bar{x} - \sigma \leq x \leq \bar{x} + \sigma$. As we increase our sample size and refine our estimates for the mean and standard deviation this becomes a more robust belief.

For the data set given in Table 2.1 for the period of the pendulum we calculate the variance $\sigma_{N-1}^2 = \dfrac{\sum d^2}{N-1}$, which numerically, from Table 2.2, is $\sigma_{N-1}^2 = \dfrac{1.58}{9} = 0.176\,\text{s}^2$. Thus using eqn (2.3) the standard deviation is $\sigma_{N-1} = 0.42\,\text{s}$, similar to the value obtained using the rough-and-ready approach.

2.4   Continuous distributions

A measurement of a physical quantity will necessarily be discrete, with the smallest division being limited by the gradations of the instrument (see Chapter 1). It is instructive to consider how the forms of the histograms of Fig. 2.1 would evolve as a function of (i) the sample size (number of data points collected), and (ii) the width of the bin used to construct the histogram. The histogram becomes smoother as more data are collected, as seen in the evolution from (a) to (d) in Fig. 2.1. Two (conflicting) factors influence the choice of bin-width; (i) there should be sufficient occurrences per bin (guaranteed by a wide bin-width), and (ii) there should be enough bins contributing to the histogram (guaranteed by a narrow bin-width). Figure 2.3 shows a histogram for 10 000 samples drawn from a normal distribution with mean 10 and standard deviation 1, constructed using different bin-widths. The distribution obviously becomes smoother as the bin-width is reduced. This trend will continue until the number of occurrences per bin is not statistically significant. It is useful to think of the limit where the bin-width tends to zero; the envelope of the histogram becomes a function which can be evaluated at any value of x. A continuous envelope is plotted for each discrete histogram in Fig. 2.3. It becomes convenient to think of a curve, or continuous distribution function, associated with the hypothetical limit of the number of data points collected, N, tending to infinity.

2.5 The normal distribution

The form of the continuous curve which is the envelope of the histogram of Fig. 2.3 has a characteristic and familiar shape. For measurements with random errors the distribution is called the **normal**, or **Gaussian**, distribution.[1] Mathematically, it is a two-parameter function:[2]

$$f(x) = \frac{1}{\sigma\sqrt{2\pi}} \exp\left[-\frac{(x-\bar{x})^2}{2\sigma^2}\right], \tag{2.4}$$

which describes the distribution of the data about the mean, $\bar{x}$, with standard deviation, σ. Figure 2.4 shows the functional form of three normalised Gaussian distributions, each with a mean of 10 and with standard deviations of 1, 2 and 3, respectively. Each curve has its peak centred on the mean, is symmetric about this value and has an area under the curves equal to 1. The larger the standard deviation, the broader the distribution, and correspondingly lower the peak value.[3]

Why is it called the normal distribution? Mathematical analysis shows that the Gaussian distribution arises as the envelope of the histogram obtained when a quantity being measured is subject to small, independent 'kicks' (or perturbations) of varying sign which contribute additively. Typically the underlying mechanisms of the perturbations are unknown. As the conditions which predict a Gaussian distribution occur often in nature, many distribution functions of natural phenomena are found to be well described by the normal distribution. We will see in Chapter 3 that the distribution of the means from a set of measurements evolves to a Gaussian shape under certain conditions.

2.6 Sample and parent distribution

When discussing distributions of experimental measurements it is important to distinguish between the **sample** and **parent** distributions. In the theory of statistics, the parent distribution refers to the number of possible measured values, ξ_i; the parent population might consist of an infinite number of values. Two independent parameters, the mean, μ, and a standard deviation, σ_{parent}, characterise the parent distribution, and are related thus:

$$\sigma_{\text{parent}} = \sqrt{\frac{\sum(\xi_i - \mu)^2}{N_{\text{parent}}}}. \tag{2.5}$$

As the mean and standard deviation are independently determined σ_{parent} is defined with N in the denominator. In practice when we take a series of measurements in an experiment, x_i, we take a selection, or sample, from this parent distribution which results in a distribution called the sample distribution. This distribution is centred on the mean of the data set, $\bar{x}$, and has a standard deviation:

$$\sigma_{\text{sample}} = \sqrt{\frac{\sum(x_i - \bar{x})^2}{N-1}}. \tag{2.6}$$

[1] This distribution is also referred to as the 'bell curve'.

[2] The prefactor $\frac{1}{\sigma\sqrt{2\pi}}$ in the expression for $f(x)$ ensures that the function is normalised: i.e. $\int_{-\infty}^{\infty} f(x)\,dx = 1$.

[3] There are many possible definitions of the 'width' of a Gaussian distribution, including Full Width at Half Maximum (FWHM), the $(1/e)$ width, the $(1/e^2)$ width. There are different conventions in different disciplines as to which to adopt; each version is proportional to the standard deviation, σ, and we will use that throughout this book.

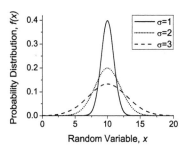

Fig. 2.4 Normalised Gaussian distributions with a mean of 10 and standard deviations of 1, 2 and 3.

*The concept of the degrees of freedom is discussed in more detail in Section 8.2.

The $(N - 1)$ is required in the denominator in the sample distribution because the mean, $\bar{x}$, is also determined from the same data set and is thus no longer independently determined and the number of degrees of freedom is one fewer.*

The goal is to make use of this sample distribution to estimate the mean and standard deviation of the parent distribution. When we conduct an experiment we discretely sample the parent distribution. In the limit that $N \to \infty$ the parent and sample distributions are the same and $\bar{x} = \mu$ and $\sigma_{\text{sample}} = \sigma_{\text{parent}}$.

If we take a single measurement we sample the parent distribution once. The most probable value of x that we would measure is the mean (see Chapter 3) and we therefore implicitly assume that $x_1 = \mu$. When we take multiple readings, however, we build up a set of values and populate the sample distribution. As we repeatedly sample the parent distribution we slowly build up a distribution of values centred on the mean of the sample distribution, $\bar{x} = \dfrac{1}{N} \sum x_i$, which becomes an increasingly better approximation of μ as N increases. As all the measurements sample the same parent distribution they are all determined with the same precision as the parent distribution, $\sigma_{\text{sample}} \approx \sigma_{\text{parent}}$. As more data are recorded the standard deviation of the data does not change, it simply becomes better defined. The evolution of the mean and standard deviation of the data shown in Fig. 2.1 is encapsulated in Table 2.3. It is clear from a visual inspection of both Fig. 2.1 and Table 2.3 that as N gets larger the mean becomes better defined, but the standard deviation hardly changes.

A key concept which we have not discussed so far is the uncertainty in the estimation of the mean. We see that as the number of data points increases the histograms become smoother, but the standard deviation does not reduce: thus the standard deviation of the sample is *not* a good measure of the error in the estimation of the mean of the parent population. We can clearly determine the position of the mean to a *better* precision than the standard deviation of the sample population. The important concept here is that of **signal-to-noise**: the precision with which we can determine the mean depends on the number of samples of the parent distribution.

Table 2.3 The evolution of the mean and standard deviation of a sample distribution with sample size, N. The parent distribution was randomly generated from a Gaussian distribution with mean $\mu = 10$, and standard deviation $\sigma_{\text{parent}} = 1$.

N	$\bar{x}$	σ_{sample}
5	9.8	0.9
10	9.5	0.7
50	10.1	0.9
100	10.0	0.9
1000	10.07	0.99

2.7   The standard error

So far we have introduced two important properties of a sequence of repeat measurements where there is scatter owing to random errors, namely the mean and standard deviation. We now introduce another crucial parameter, the uncertainty in the location of the mean.

Figure 2.5 shows histograms of a simulation where 2500 points are chosen from a normal parent distribution with mean $\mu = 10$ and standard deviation $\sigma_{\text{parent}} = 1$. In part (a) the histogram of the measurements is normal, centred as expected on 10, and has a standard deviation of $\sigma_{\text{sample}} = 1$. The upper part of the panel shows that most measurements are within two standard deviations of the mean of the parent distribution, with very few points with a deviation larger in magnitude than two standard deviations. For part (b) the same data set was partitioned differently. The mean of every five points was calculated,

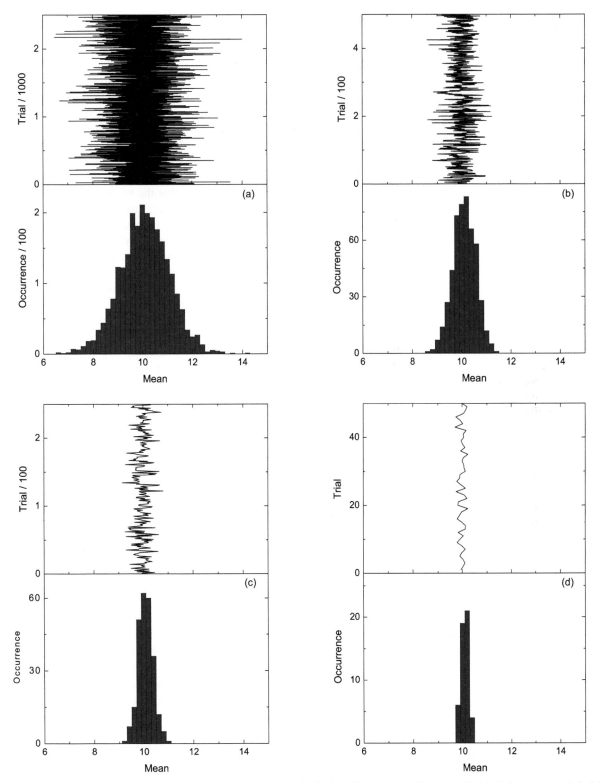

Fig. 2.5 Histograms of 2500 data points chosen from a normal parent distribution with mean $\mu = 10$ and standard deviation $\sigma_{parent} = 1$. In (a) the raw data are plotted, in (b) a mean of every five data points was calculated, and these 500 means and their distribution are plotted. In (c) 10 points were used to calculate 250 means, and in (d) 50 points were averaged to generate 50 means. Averaging over more points greatly reduces the statistical fluctuations, and reduces the width of the histograms. Each histogram has the same bin-width. The mean, $\bar{x}$, and standard deviation, σ_{sample}, are (a)10.0, 1.0, (b) 10.0, 0.5, (c) 10.0, 0.3 and (d) 10.00, 0.14.

yielding 500 mean values; the histogram shows the distribution of these means. It is evident from the width of the histogram that the distribution of these 500 means is significantly narrower than the distribution of the unaveraged data. Averaging five data points greatly reduces the statistical fluctuations in the distribution of the means. This trend continues in part (c) where the data set was partitioned into 250 measurements of means of 10 points, and in part (d) where 50 measurements of means obtained from 50 data points are plotted.

The width of the histogram of means is a measure of the precision of the mean. It is clearly evident from Fig. 2.5 that the width of the histogram of the means decreases as the size of the sample used to calculate the mean increases—this is a consequence of averaging over statistical fluctuations. The width of the histogram of means is the **standard deviation of the mean**, also known as the **standard error**, α. When the number of measurements involved in calculating the mean increases, the means are better defined; consequently the histograms of the distributions of the means are narrower. Note that the precision with which the mean can be determined is related to the number of measurements used to calculate the mean. In practice one does not generate histograms of the means based on trials containing many measurements; rather one uses all N measurements x_i to calculate one value for the mean $\bar{x}$. We expect the standard error (the standard deviation of the mean), α, to decrease as the number of data points we collect, N, increases. In Chapter 4 we discuss ways of combining measurements of the same quantity, and we show[4] that the standard error is reduced by a factor of $\sqrt{N}$ with respect to the sample standard deviation:

[4]This is a special case of the **weighted mean**—see Chapter 4.

$$\alpha = \frac{\sigma_{N-1}}{\sqrt{N}}. \tag{2.7}$$

A data set containing N multiple readings yields one value of the mean. Thus we should quote our findings as the mean ± the error on the mean, i.e. $\bar{x} \pm \alpha = \bar{x} \pm \dfrac{\sigma_{N-1}}{\sqrt{N}}$. In other words, we are saying that there is a two-thirds chance that the measured parameter is within the range $\bar{x} - \alpha \leq x \leq \bar{x} + \alpha$. One can interpret the standard error as being a standard deviation, not of the measurements, but rather the mean: this is why the standard error is also called the standard deviation of the mean (**SDOM**).[5]

[5]The reduction of the standard error by a factor of $\sqrt{N}$ with respect to the standard deviation of the measurements will reappear in the discussion of the central limit theorem in Chapter 3.

The reduction of the standard deviation of the mean with respect to the standard deviation of the parent population is inherent in Fig. 2.5. The standard deviation of the mean based on using five measurements to calculate the mean is $\alpha = 0.47$, when based on 10 the standard deviation of means is $\alpha = 0.30$, and, finally, when based on 50 the standard deviation of means is $\alpha = 0.14$. From eqn (2.7) we would expect the error in the mean to be $\sigma_{\bar{x}} = \dfrac{\sigma_{\text{parent}}}{\sqrt{N}} = 0.45, 0.32, 0.14$, in excellent agreement with our findings.[6]

[6]We keep two significant figures to illustrate the argument here, although with so few data points we would generally only quote the standard deviation to one significant figure.

2.7.1   The error in the error

There is one last statistical quantity which we must consider before we unveil the procedure for how to report the best value, and its uncertainty, for a sequence of repeat measurements. Given that the precision with which we know the mean varies with the number of measurements, we need to ensure that we present the results in a systematic manner which reflects our confidence in the error: i.e. we need to quantify the error in the error.

There exists a formula for the fractional error in the error (Squires 2001, Appendix B); it is defined as

$$\text{error in the error} = \frac{1}{\sqrt{2N - 2}}, \qquad (2.8)$$

and plotted in Fig. 2.6. It should be noted that the error on the error is a slowly decreasing function with respect to N. For example, with only five measurements the error estimate is only good to 1 part in 3 (35%). As the sample size increases, the error in the error decreases, and we can be more confident in our results, allowing for more significant figures to be quoted. Note that the error in the error does not fall to a few percent (allowing two significant figures to be quoted meaningfully) until approximately 10 000 data points have been collected (see Exercise 2.4). Conversely, care should be taken when choosing the number of appropriate significant figures if the first significant figure of the error is 1—rounding an error of 1.4 to 1, or 1.51 to 2 causes a change in the error of approximately 25%. The following rule is generally adopted:[7]

> Quote the error to one significant figure.

Note that it is extremely rare to quote errors to three significant figures or higher.[8] As we saw in Section 1.1, even the currently accepted values for the fundamental constants have their errors quoted only to two significant figures. Note also that there is no rule about how many significant figures are included in the mean—this is ascertained *after* the error (and its error) are evaluated. The value of the acceleration due to gravity deduced in the worked example after 7 500 measurements has an error known to two significant figures, and a mean known to four significant figures, whereas Avogadro's number has an error known to two significant figures, and a mean known to nine significant figures (see Section 1.1).

2.8 Reporting results

From the preceding section, we can formulate the following procedures to be considered when we quote our results:

(1) Analyse the experimental data and calculate the mean; keep all significant figures at this stage.
(2) Calculate the standard error (the error in the mean); keep all significant figures at this stage.
(3) Think about how many significant figures should be retained for the error having reflected on the number of data points collected.
(4) Round the mean to the appropriate decimal place.

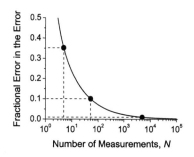

Fig. 2.6 The fractional error in the error is plotted as a function of the number of measurements, N. Note the logarithmic scale for the abscissa. For five measurements the fractional uncertainty in the error is 35%; 50 measurements are needed for the error to be known to 10%, and 5000 measurements to achieve a 1% fractional uncertainty.

[7]**Corollary** (i) If you have collected approximately 10 000 data points, or more, consider quoting the error to two significant figures; (ii) if the first significant figure of the error is 1, consider quoting the second significant figure.

[8]**Worked example** Analysis of repeat measurements of the acceleration due to gravity, g, yields $\bar{g} = 9.812\,3456$ m s^{-2}, with $\alpha = 0.032\,1987$ m s^{-2}.

- If this answer was based on 10 measurements, you would report $g = (9.81 \pm 0.03)$ m s^{-2};
- if this answer was based on 7 500 measurements, you would consider reporting $g = (9.812 \pm 0.032)$ m s^{-2}.

If another measurement technique has results which are $\bar{g} = 9.817\,654$ m s^{-2} and $\alpha = 0.101\,23$ m s^{-2}, then

- If this answer was based on 10 measurements, you would report $g = (9.8 \pm 0.1)$ m s^{-2};
- if this answer was based on 500 measurements, you would consider reporting $g = (9.82 \pm 0.10)$ m s^{-2}.

Returning to the data of Table 2.1, the mean is $\bar{T} = 9.9$ s, standard deviation $\sigma_{N-1} = 0.42$ s, and standard error $\alpha = \dfrac{\sigma_{N-1}}{\sqrt{N}} = 0.13$ s. As the analysis is based on so few data points only the first significant figure of the error is retained; the result is reported as $T = (9.9 \pm 0.1)$ s.

2.8.1   Rounding and significant figures

RULE OF THUMB: If an error is not quoted assume that the uncertainty is in the last reported digit.

The theme of this chapter has been that all measurements are subject to uncertainty. A working rule is that, in the absence of an error being quoted, we assume that a number has significance equal to a single unit in the last figure quoted. Thus if we were to say that the resistance of a resistor was 97 Ω, it is said to have an absolute uncertainty of 1 Ω; a resistor with a value of 100.04 Ω indicates an absolute uncertainty of 0.01 Ω. The former value is said to be known to two significant figures, the latter to five. Confusion can occur in ascertaining how many significant figures a number has when zeroes are involved.

Rules for identifying significant digits

- All non-zero digits are significant:
  2.998×10^8 m s^{-1} has four significant figures.
- All zeroes between non-zero digits are significant:
  $6.022\,141\,79 \times 10^{23}$ mol^{-1} has nine significant figures.
- Zeroes to the left of the first non-zero digits are not significant: 0.51 MeV has two significant figures.
- Zeroes at the end of a number to the right of the decimal point are significant: 1.60×10^{-19} C has three significant figures.
- If a number ends in zeroes without a decimal point, the zeroes might be significant: 270 Ω might have two or three significant figures.

The ambiguity in the last rule can be resolved by the use of so-called **scientific notation**. For example, depending on whether two or three significant figures is appropriate, we could write 270 Ω as 0.27 kΩ, or 2.7×10^2 Ω, both of which have two significant figures; or 0.270 kΩ, or 2.70×10^2 Ω, both of which have three significant figures. Note that the entries 0.3 kΩ and 300 Ω in a lab book carry very different significance.

RULE OF THUMB: To avoid confusion when numbers end in zeros, report your values using **scientific notation.**

Significant figures must also be considered when carrying out calculations. It is important to carry all digits through to the final result before rounding to avoid **rounding errors** which compromise the accuracy of the final result. The principle is the following:

> The precision of a calculated result is limited by the least precise measurement in the calculation.

Rules for rounding to the appropriate number of significant figures

Decide which is the last digit to keep, then:

- Leave the last digit unchanged if the next digit is 4 or lower: 6.62×10^{-34} becomes 6.6×10^{-34} if only two significant figures are appropriate.
- Increase the last digit by 1 if the next digit is 6 or higher: 5.67×10^{-8} becomes 5.7×10^{-8} if only two significant figures are appropriate.

If the digit after the last one to be retained is 5 the recommended procedure is to choose the even round value.[9]

- Leave the last digit unchanged if it is even. For example: 3.45 becomes 3.4 if only two significant figures are appropriate.
- Increase the last digit by 1 if it is odd. For example: 3.55 becomes 3.6 if only two significant figures are appropriate.

[9]This round-to-even method avoids bias in rounding, because half of the time we round up, and half of the time we round down.

In addition and subtraction, the result is rounded off to the same number of decimal places as the number with the least number of decimal places. For example, $1.23 + 45.6$ should be quoted as 46.8. This reflects the fact that we do not know whether the 45.6 is 45.56 or 45.64 to the next decimal place.

In multiplication and division, the answer should be given to the same number of significant figures as the component with the least number of significant figures. For example, 1.2×345.6 is evaluated as 414.72 but quoted as 4.1×10^2 on account of the least precise value having only two significant figures.

It is important to carry all significant figures through long calculations to avoid unnecessary rounding errors. Rounding to the appropriate precision should only be done at the end of the calculation.

There are some **exact numbers** which can be considered to have an infinite number of significant figures, and they do not influence the precision to which a result is quoted. They are often found in conversion factors (such as π or $\sqrt{2}$) and when counting: there are exactly 100 centimetres in 1 metre; there are 14 students in the laboratory.

2.9   The five golden rules

We finish this chapter with the five golden rules which must be obeyed when reporting a parameter which was determined experimentally.

(1) The best estimate of a parameter is the mean.
(2) The error is the standard error in the mean.
(3) Round up the error to the appropriate number of significant figures.
(4) Match the number of decimal places in the mean to the standard error.
(5) Include units.

Chapter summary

- The presence of random uncertainties can be ascertained by taking repeat measurements.
- For N measurements $x_1, x_2, \ldots, x_N$ the mean, $\bar{x}$, is the best estimate of the quantity x.

- The standard deviation of the sample σ_{N-1} gives a measure of the precision of the measurements—two-thirds of the measurements will lie within σ_{N-1} of the mean.
- The uncertainty in the location of the centre of the distribution is given by α, the standard error of the mean. The error decreases (slowly) with more measurements:

$$\alpha = \frac{\sigma_{N-1}}{\sqrt{N}}.$$

- The result of repeated measurements is reported as $\bar{x} \pm \alpha$.
- The fractional error in the error decreases very slowly with increasing the number of measurements, hence the error is usually quoted to only one significant figure.

Exercises

(2.1) *Mean, standard deviation and standard error (1)*

An experiment was conducted to determine the concentration of a sodium hydroxide solution. The eight repeat measurements of the volume of hydrochloric acid titrated (all in ml) are: 25.8, 26.2, 26.0, 26.5, 25.8, 26.1, 25.8 and 26.3. Calculate (i) the mean, (ii) the standard deviation using the rough-and-ready approach; (iii) the standard deviation using eqn (2.3); (iv) the standard error of the volume.

(2.2) *Mean, standard deviation and standard error (2)*

12 measurements of the sensitivity of a photodiode circuit (in amps/watt) are: 5.33, 4.95, 4.93, 5.08, 4.95, 4.96, 5.02, 4.99, 5.24, 5.25, 5.23 and 5.01. Calculate (i) the mean, (ii) the standard deviation using eqn (2.3); (iii) the standard error.

(2.3) *Reduction of the standard error*

In a magnetometry experiment, after a minute of collecting data the statistical noise was reduced to 1 picotesla. For how much longer should data be collected in order to reduce the random error by a factor of 10?

(2.4) *Error in the error*

Consider a set of measurements with the standard error calculated to be $\alpha = 0.987\,654\,321$. Here we address the question of how many significant figures should be quoted. Construct a spreadsheet with four columns. The first column should be N, the number of measurements on which α is based. In the second column write α to the nine significant figures quoted above. The third

and fourth columns should be $\alpha \times \left(1 - \dfrac{1}{\sqrt{2N-2}}\right)$ and $\alpha \times \left(1 + \dfrac{1}{\sqrt{2N-2}}\right)$, respectively. As we are interested in the variation over a large dynamic range, choose values for N such as 2, 3, 5, 10, 20, 30, etc. Verify the statement from Section 2.7.1 that the number of data points, N, needs to approach a few tens of thousands before the second significant figure in the error can be quoted, i.e. when the values in the three columns become equal to the second significant figure. Repeat the analysis for the case where $\alpha = 0.123\,456\,789$, i.e. the first significant digit of the error is 1. How many data points must be collected before the third significant figure can be quoted?

(2.5) *Reporting results (1)*

Fifteen measurements of a resistance are quoted here, based on approximately 10 repeat measurements. Only three of them obey the five golden rules. Identify the mistakes in the other results.

(i) $(99.8 \pm 0.270) \times 10^3\ \Omega$,

(ii) $(100 \pm 0.3) \times 10^3\ \Omega$,

(iii) $(100.0 \pm 0.3) \times 10^3\ \Omega$,

(iv) $(100.1 \pm 0.3) \times 10^3$,

(v) $97.1 \times 10^3 \pm 276\ \Omega$,

(vi) $(99.8645 \pm 0.2701) \times 10^3\ \Omega$,

(vii) $98.6 \times 10^3 \pm 3 \times 10^2\ \Omega$,

(viii) $99.4 \times 10^3 \pm 36.0 \times 10^2 \ \Omega$,

(ix) $101.5 \times 10^3 \pm 0.3 \times 10^1 \ \Omega$,

(x) $(99.8 \pm 0.3) \times 10^3 \ \Omega$,

(xi) $95.2 \times 10^3 \pm 273 \ \Omega$,

(xii) $98,714 \pm 378 \ \Omega$,

(xiii) $99000 \pm 278 \ \Omega$,

(xiv) $98,714 \pm 3 \times 10^3 \ \Omega$,

(xv) $98900 \pm 300 \ \Omega$.

(2.6) *Reporting results (2)*

Analysis of a Rydberg spectrum yields a quantum defect, δ, for each line. How would you report the results if you obtain (i) $\bar{\delta} = 3.273\,46$, with $\alpha_\delta = 0.019\,13$ from five measurements; (ii) $\bar{\delta} = 3.265\,13$, with $\alpha_\delta = 0.002\,506$ from 50 measurements; or (iii) $\bar{\delta} = 3.266\,81$, with $\alpha_\delta = 0.000\,270$ from 100 measurements?

(2.7) *Significant figures*

Round up the following numbers to (a) two significant figures, and (b) four significant figures:

(i) 602.20,

(ii) 0.001\,3806,

(iii) 0.022\,413\,83,

(iv) 1.602\,19,

(v) 91.095,

(vi) 0.1660,

(vii) 299\,790\,000,

(viii) 66.2617,

(ix) 0.000\,006\,672 and

(x) 3.141\,593.

(2.8) *Scientific notation*

Rewrite the ten numbers from Exercise (2.7) in scientific notation.

(2.9) *Superfluous precision*

A car covers a distance of 250 m in 13 s; the average speed is calculated to the 10 decimal places of the calculator as $19.230\,769\,23 \ \text{m s}^{-1}$. Explain why it is incorrect to believe all of the significant figures of the quoted speed.

Uncertainties as probabilities

In this chapter we develop further some of the ideas from Chapter 2, and consider the link between uncertainties in measurements and probabilities. The discrete histograms considered so far will be augmented by the concept of a probability distribution function. The most important distribution function for error analysis is a Gaussian, and we extend the discussion from Chapter 2 about the pertinent properties of this function. We are then able to discuss the confidence limits in error analysis, our confidence that the accepted value of a measured quantity will lie within a certain range. We also consider the converse, namely what is the range of a variable within which a certain percentage of measurements is likely to lie.

When counting discrete random events, such as the number of particles emitted from a radioactive source, it transpires that the distribution does not follow a Gaussian distribution as discussed so far, but rather is better described by a Poisson distribution. We discuss the relevant features of this distribution for error analysis.

3.1 Distributions and probability

In Section 2.4 we introduced for a discrete histogram the concept of the envelope curve, or continuous distribution function, associated with the hypothetical limit of the number of data points collected, N, tending to infinity. In statistics a continuous random variable, say x, has a probability distribution which may be specified in terms of a probability distribution function (or probability density function), $P_{DF}(x)$. A probability distribution function has the following properties:

(1) The distribution is said to be **normalised**, or proper, if

$$\int_{-\infty}^{\infty} P_{DF}(x)\,\mathrm{d}x = 1. \tag{3.1}$$

(2) The probability that x lies between two values x_1 and x_2, with $x_1 \leq x_2$, is

$$P(x_1 \leq x \leq x_2) = \int_{x_1}^{x_2} P_{DF}(x)\,\mathrm{d}x. \tag{3.2}$$

(3) The expectation of the nth power of the random variable x is

$$\overline{x^n} = \int_{-\infty}^{\infty} P_{\text{DF}}(x) \, x^n \, dx. \tag{3.3}$$

The mean can be calculated by applying eqn (3.3) with $n = 1$:

$$\bar{x} = \int_{-\infty}^{\infty} P_{\text{DF}}(x) \, x \, dx. \tag{3.4}$$

The variance, σ^2, is defined as:

$$\sigma^2 = \int_{-\infty}^{\infty} P_{\text{DF}}(x) \, (x - \bar{x})^2 \, dx = \int_{-\infty}^{\infty} P_{\text{DF}}(x) \left(x^2 + \bar{x}^2 - 2x\bar{x} \right) dx. \tag{3.5}$$

By applying eqn (3.3) with $n = 2$ and eqn (3.4) we obtain

$$\sigma^2 = \overline{x^2} - \bar{x}^2, \tag{3.6}$$

where $\overline{x^2}$ is the mean of the squares of the random variable x.

3.2   The Gaussian probability distribution function

As we discussed in Section 2.5 the most important function in error analysis is the Gaussian (or normal) probability density distribution. For the sake of brevity we usually refer to it as the Gaussian probability distribution, or simply the Gaussian distribution. In this chapter we will write the function as $G(x; \bar{x}, \sigma)$, where:

$$G(x; \bar{x}, \sigma) = \frac{1}{\sigma\sqrt{2\pi}} \exp\left[-\frac{(x - \bar{x})^2}{2\sigma^2} \right], \tag{3.7}$$

to emphasise that the function has x as a variable, and has the mean, $\bar{x}$, and standard deviation, σ, as two parameters.

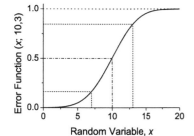

3.2.1   Probability calculations

We can use eqn (3.7) to determine the fraction of the data which is expected to lie within certain limits. For a Gaussian the cumulative probability is the well-known error function $\text{Erf}(x_1; \bar{x}, \sigma)$:

$$\text{Erf}(x_1; \bar{x}, \sigma) = \int_{-\infty}^{x_1} G(x; \bar{x}, \sigma) \, dx. \tag{3.8}$$

This function has the mean, $\bar{x}$, and the standard deviation, σ, as parameters, and is evaluated at $x = x_1$. The error function is plotted in Fig. 3.1 and tabulated in

Fig. 3.1 The error function of the random variable x is the cumulative integral (the area under the curve) of a Gaussian from $-\infty$ to x. Here it is plotted for a Gaussian with mean $\bar{x} = 10$ and standard deviation $\sigma = 3$. The function is antisymmetric about the mean; is equal to 0.159 for $x = \bar{x} - \sigma$, 0.500 for $x = \bar{x}$, and 0.841 for $x = \bar{x} + \sigma$; and the error function tends asymptotically to 1.

many data-analysis packages. The fractional area under the curve between the bounds x_1 and x_2 is thus:

$$P(x_1 \leq x \leq x_2) = \frac{1}{\sigma\sqrt{2\pi}} \int_{x_1}^{x_2} \exp\left[-\frac{(x - \bar{x})^2}{2\sigma^2}\right] dx$$

$$= \mathrm{Erf}(x_2; \bar{x}, \sigma) - \mathrm{Erf}(x_1; \bar{x}, \sigma). \tag{3.9}$$

Figure 3.2 shows the relationship between the error function and the area under certain portions of the Gaussian distribution function. For a Gaussian with mean $\bar{x} = 10$ and standard deviation $\sigma = 3$, what fraction of the data lies in the interval $5 \leq x \leq 11.5$? For this Gaussian the probability of obtaining a value of $x \leq 11.5$ is 0.69, and the probability of obtaining a value of $x \leq 5$ is 0.05; hence 64% of the area under the curve is in the interval $5 \leq x \leq 11.5$.

3.2.2   Worked example—the error function

A box contains 100 Ω resistors which are known to have a standard deviation of 2 Ω. What is the probability of selecting a resistor with a value of 95 Ω or less? What is the probability of finding a resistor in the range 99–101 Ω?

Let x represent the value of the resistance which has a mean of $\bar{x} = 100\,\Omega$. The standard deviation is given as $\sigma = 2\,\Omega$. The probability of selecting a resistor with a value of 95 Ω or less can be evaluated from eqn (3.8):

$$P = \mathrm{Erf}(95; 100, 2) = 0.0062.$$

Applying eqn (3.9) we find the probability of finding a resistor in the range 99–101 Ω

$$P = \mathrm{Erf}(101; 100, 2) - \mathrm{Erf}(99; 100, 2) = 0.38.$$

These integrals can be evaluated numerically, found in look-up tables or found by using appropriate commands in spreadsheet software.

3.3   Confidence limits and error bars

Consider first the fraction of the data expected to lie within one standard deviation of the mean:

$$P = \frac{1}{\sigma\sqrt{2\pi}} \int_{\bar{x}-\sigma}^{\bar{x}+\sigma} \exp\left[-\frac{(x - \bar{x})^2}{2\sigma^2}\right] dx$$

$$= \mathrm{Erf}(\bar{x} + \sigma; \bar{x}, \sigma) - \mathrm{Erf}(\bar{x} - \sigma; \bar{x}, \sigma). \tag{3.10}$$

Numerically, this integral is equal to 0.683 (to three significant figures). In other words, approximately two-thirds of the total area under the curve is within a standard deviation of the mean. This is the origin of the two-thirds terms used extensively in Chapter 2. We can now quantify our statements about the standard deviation of a sample of measurements: we are confident, at the 68% level, that, were we to take another measurement, the value would lie within one standard deviation of the mean.

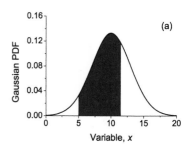

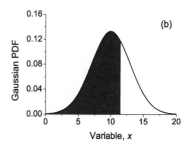

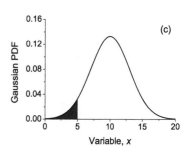

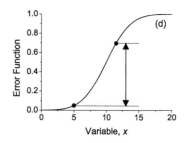

Fig. 3.2 A Gaussian with mean $\bar{x} = 10$ and standard deviation $\sigma = 3$ is shown. In (a) the fraction of the curve within the interval $5 \leq x \leq 11.5$ is shaded; this is equal to the difference between (b) the error function evaluated at 11.5 and (c) the error function evaluated at 5. The relevant values of the error function are highlighted in (d).

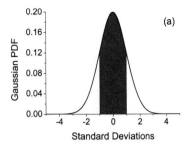

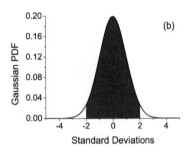

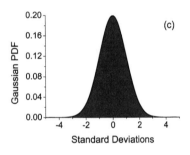

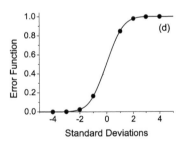

Fig. 3.3 The shaded areas of the Gaussian curves show the fraction of data within (a) one standard deviation of the mean, (b) two standard deviations, and (c) three standard deviations. The corresponding points on the error function, along with those for four standard deviations, are indicated in (d).

3.3.1 Extended ranges

By evaluating the error function of eqn (3.9) we can define the probabilities that the data lie within an interval defined by two, three, etc. standard deviations from the mean; these values are indicated in Fig. 3.3, and tabulated in Table 3.1.

Table 3.1 The fraction of the data which lies within different ranges of a Gaussian probability distribution function.

Centred on mean	$\pm\sigma$	$\pm1.65\sigma$	$\pm2\sigma$	$\pm2.58\sigma$	$\pm3\sigma$
Measurements within range	68%	90%	95%	99.0%	99.7%
Measurements outside range	32%	10%	5%	1.0%	0.3%
	1 in 3	1 in 10	1 in 20	1 in 100	1 in 400

Recalling the interpretation of the standard error as a standard deviation of the mean we can also calculate other confidence intervals. Whereas often in the physical sciences the error bar is taken as one standard deviation of the mean (the standard error), other conventions exist; in other disciplines the 95% confidence limit is often used. Evaluating the error function of eqn (3.9) it can be shown that 95.0% of the measurements lie within the range $\pm1.96\sigma$. Therefore if a data set of N measurements has a mean $\bar{x}$ and standard deviation σ_{N-1}, we would report the result at the 95% confidence limit as $\bar{x} \pm 1.96 \times \dfrac{\sigma_{N-1}}{\sqrt{N}}$. Different confidence limits can be used by scaling the standard error appropriately.

It should be noted that in the above discussion it is assumed that the standard deviation of the Gaussian distribution is precisely and accurately known. When σ_{N-1} is ascertained from experimental data, especially from a small number of repeat measurements, greater care is needed with confidence limits. In Chapter 8 we will discuss the Student's t distribution which is more appropriate for interval estimation from a small number of data points.

3.3.2 Rejecting outliers

Here we discuss a controversial topic in data analysis, that of rejecting outliers. From Table 3.1 we learn that we should not be very surprised if a measurement is in disagreement with the accepted value by more then one error bar, α; indeed, for a group of 15 students in a laboratory performing the same experiment, we would expect approximately five to report results where the magnitude of the difference between the accepted and measured results is greater than one error bar. However, as the fractional area under a Gaussian curve beyond 3σ or 5σ is only 0.3% and 6×10^{-5}%, respectively, we expect such large deviations to occur very infrequently.

Chauvenet's criterion[1] (Taylor 1997, Section 6.2; Bevington and Robinson 2003 p. 56) is a test based on the Gaussian distribution with the aim of assessing whether one data point which lies many error bars from the mean (an outlier) should be regarded as spurious and hence discarded. The criterion is equivalent to the statement 'a data point is rejected from a sample if the number of events we expect to be farther from the mean than the suspect point, for the sample's mean and standard deviation, is less that a half'. The procedure is as follows:

(1) For your N measurements $x_1, x_2, \ldots, x_N$, calculate the mean, $\bar{x}$, and standard deviation, σ_{N-1}.

(2) For the potential outlier, x_{out}, use the error function to find the probability that a result would randomly differ from the mean by the same amount, or more:

$$P_{out} = 1 - P\left(\bar{x} - x_{out} \leq x \leq \bar{x} + x_{out}\right)$$
$$= 1 - \left[\mathrm{Erf}\left(\bar{x} + x_{out}; \bar{x}, \sigma\right) - \mathrm{Erf}\left(\bar{x} - x_{out}; \bar{x}, \sigma\right)\right].$$

(3) Multiply the probability of there being such an outlier with the number of data points, $n_{out} = P_{out} \times N$.

(4) If the number n_{out} is less than one-half, then Chauvenet's criterion states that you reject the outlier x_{out}. One then recalculates the mean and standard deviation for the remaining $N - 1$ data points.

This controversial procedure should be applied with care. One should always ask if there is a possible reason why the outlier occurred. We are assuming that the data follow a Gaussian distribution, therefore a whole class of potentially interesting questions about the form of the distribution of measurements from a particular experiment would be severely compromised by applying Chauvenet's criterion. Rejecting outliers is easier to justify when a model of the expected distribution of the measured variable is known from previous experiments or a theoretical prediction. If there is a limited set of data (for example, you only had access to the telescope for one night's observation) consider removing outliers. A better strategy is to repeat the measurement, revisiting the settings which produced the outlier if it is reasonably straightforward to do so.

3.3.3 Experimental example of a Gaussian distribution

Figure 3.4 shows the signal output from a photodiode as a function of time, and in part (b) a histogram of the distribution of data. The mean and standard deviation of the data were calculated, and part (b) also has a Gaussian with the same mean and standard deviation superimposed on the the data.

[1] Consider 10 data points, where a suspected outlier is three standard deviations removed from the mean. Should this data point be retained? From Table 3.1 we see that $P_{out} = 0.003$, thus the expected number of outliers is 0.03. As this is much less than 0.5 Chauvenet's criterion would indicate that the suspect data point be rejected. The mean, standard deviation and standard error of the remaining nine data points should then be recalculated.

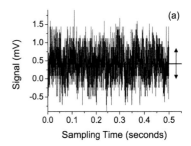

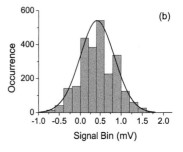

Fig. 3.4 Part (a) displays the voltage across a photodiode as a function of time sampled 2 500 times. In (a) the standard deviation σ is indicated by the arrow centred on the mean of the signal (0.4 mV). In (b) the histogram of the number of occurrences in 0.2 mV bins is indicated, together with a smooth curve. The curve is the Gaussian distribution which has the same mean and standard deviation as the data. A comparison of what fraction lies within certain bounds of the mean is encapsulated in Table 3.2.

Table 3.2 A comparison between experimental noise and a Gaussian model.

Centred on mean	$\pm\sigma$	$\pm1.65\sigma$	$\pm2\sigma$	$\pm2.58\sigma$	$\pm3\sigma$
Expected values	68%	90%	95%	99.0%	99.7%
Fraction of data points within range	67%	89%	95%	99.3%	99.9%

We can compare the percentage of data points that fall within one, two, etc. standard deviations of the mean, and compare with the values expected from a Gaussian distribution. This is shown in Table 3.2 and the agreement between theory and experiment is excellent.

The fraction of data lying within these bounds is very close to that expected for a Gaussian distribution. In Chapter 8 we present techniques which allow us to answer more quantitatively the question 'how good a fit to the data is the Gaussian distribution?'.

3.3.4   Comparing experimental results with an accepted value

If we are comparing an experimentally determined value with an accepted value, we can use Table 3.1 to define the likelihood of our measurement being accurate. The procedure adopted involves measuring the discrepancy between the experimentally determined result and the accepted value, divided by the experimentally determined standard error. If your experimental result and the accepted value differ by:

- up to one standard error, they are in **excellent agreement**;
- between one and two standard errors, they are in **reasonable agreement**;
- more than three standard errors, they are in **disagreement**.

3.4   Poisson probability function for discrete events

The Poisson distribution is the distribution function appropriate to modelling discrete events. It expresses the probability of a number of relatively *rare events* occurring in a fixed time if these events occur with a known average rate, and are independent of the time since the last event. The conditions under which a Poisson distribution holds are when (Bevington and Robinson 2003, p. 23):

- counts are of rare events;
- all events are independent;
- the average rate does not change over the period of interest.

One frequently encountered example of these conditions arises when dealing with counting—especially radioactive decay, or photon counting using a

Geiger tube. Unlike the normal distribution the only parameter we need to define is the average count in a given time, $\overline{N}$. The average count is the product of the average count rate, λ, and the time for which we count, τ: $\overline{N} = \lambda \tau$. For example, in a particular radioactivity experiment the average count rate is $\lambda = 1.5 \text{ s}^{-1}$, and data are collected for 10 s. The average count expected is thus 15. Now, for different repeats of the experiment the random character of radioactive decay will mean there is a fluctuation in the number, N, of actual counts registered. The Poisson probability distribution is the single-parameter function which allows us to find the probability that there are exactly N occurrences (N being a non-negative integer, $N = 0, 1, 2, \ldots$) given the average count, $\overline{N}$. Note that the average count does not have to be an integer: if counts are only collected for 1 s for this example we would expect a mean count of $\overline{N} = 1.5$. The Poisson probability distribution is defined as:

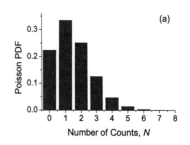

$$P\left(N; \overline{N}\right) = \frac{\exp\left(-\overline{N}\right) \overline{N}^N}{N!}. \tag{3.11}$$

The denominator is the factorial function, and is defined such that $N! = N \times (N-1) \times \cdots \times 3 \times 2 \times 1$. As an example, $4! = 4 \times 3 \times 2 \times 1 = 24$. The functional form of eqn (3.11) is shown in Fig. 3.5 for two Poisson distributions, with means $\overline{N} = 1.5$ and $\overline{N} = 15$, respectively. Note that the Poisson distribution is only defined at integer values of N, i.e. there is some finite probability of having zero counts in the time τ (0.223 for $\overline{N} = 1.5$), a probability of 0.335 of obtaining one count, etc. Each distribution is peaked close to the average value, $\overline{N}$; is *asymmetric* about this value; and, in common with all proper probability functions, the sum of the probabilities is equal to 1.

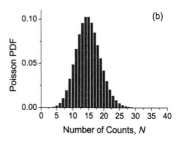

Fig. 3.5 Poisson distributions with (a) mean $\overline{N} = 1.5$ and (b) mean $\overline{N} = 15$. The probability of occurrence is plotted against the number of counts, N. Note the asymmetry of the distribution about the mean.

We can find the mean and the standard deviation of the Poisson probability function using the discrete equivalent to eqn (3.3):

$$\overline{N} = \sum P\left(N; \overline{N}\right) N, \tag{3.12}$$

and

$$\overline{N^2} = \sum P\left(N; \overline{N}\right) N^2. \tag{3.13}$$

Evaluating the summations in eqns (3.12) and (3.13) we find that the average count (unsurprisingly) is $\overline{N}$, and the standard deviation of a Poisson distribution is simply $\sigma = \sqrt{\overline{N}} = \sqrt{\lambda \tau}$. It is worth emphasising again that, in contrast to the Gaussian distribution, only a single parameter $(\overline{N})$ is needed to specify a Poisson distribution—the mean and standard deviation are not independent.[2]

[2]Most spreadsheet software has built-in functions for evaluating the Poisson distribution.

3.4.1 Worked example—Poisson counts

A safety procedure at a nuclear power plant stops the nuclear reactions in the core if the background radiation level exceeds 13 counts per minute. In a random sample, the total number of counts recorded in 10 hours was 1980. What is the count rate per minute and its error? What is the probability that during a random one-minute interval 13 counts will be recorded? What is the probability that the safety system will trip?

The number of counts recorded in a minute will follow a Poisson distribution. The mean count rate is $\lambda = 1980/(10 \times 60) = 3.30$ counts per minute. The error in the number of counts is $\sqrt{1980} = 44.5$; therefore the error in the count rate is $\alpha_\lambda = \sqrt{1980}/(10 \times 60) = 0.07$ counts per minute. The probability of having 13 counts in a minute is given by the Poisson distribution, with $\overline{N} = \lambda \times \tau = 3.30$ and $N = 13$:

$$P(N = 13; 3.3) = \frac{\exp(-3.3)\, 3.3^{13}}{13!} = 3.3 \times 10^{-5}.$$

To calculate the probability of having 13 or more counts it is, in fact, easier to evaluate the probability of detecting 12 counts or fewer; these numbers are complementary.

$$P(N \geq 13; 3.3) = 1 - [P(0; 3.3) + P(1; 3.3) + \cdots P(12; 3.3)]$$
$$= 4.2 \times 10^{-5}.$$

Therefore, based on the number of counts recorded in a minute, the probability that the safety system will trip is 4.2×10^{-5}.

3.4.2   Error bars and confidence limits for Poisson statistics

For experiments that involve counting rare independent events with a constant average rate, the Poisson distribution is used.

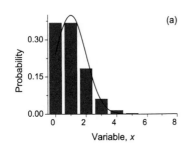

(a)

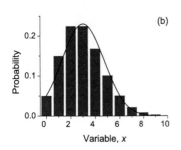

(b)

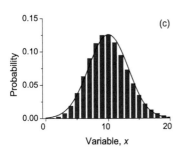

(c)

> If an experiment yields a mean count of N the best estimate of the error in this quantity is $\sqrt{N}$. We therefore report the measurement as $N \pm \sqrt{N}$.

For Poisson counts the **fractional uncertainty** is $\dfrac{\sqrt{N}}{N} = \dfrac{1}{\sqrt{N}}$. For situations where the number of counts expected is very low (a weak source, an inefficient detector or a short data collection interval) the Poissonian fluctuations in the random events lead to a poor signal-to-noise ratio. This phenomenon was referred to as shot noise in Chapter 1.

Note also the interpretation of the error bar as the standard deviation; having measured N counts in a given time, we believe there is a two-thirds chance that another run of the experiment will yield a count in the interval $N - \sqrt{N}$ to $N + \sqrt{N}$. Owing to the asymmetry of the Poisson distribution one should not apply blindly the confidence limits for a Gaussian distribution from Table 3.1.

3.4.3   Approximations for high means

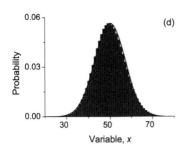

(d)

Fig. 3.6 The evolution of the Poisson distribution as the mean increases. The Gaussian distribution as a function of the continuous variable x is superimposed. The mean of the distribution is (a) 1, (b) 3, (c) 10 and (d) 50.

A feature of the Poisson distribution which distinguishes it from a Gaussian is the asymmetric distribution, most prominent for a low mean. However, as the mean gets larger, the Poisson distribution gets more symmetric, and closely resembles a Gaussian. Figure 3.6 shows the evolution of the (discrete) Poisson function compared with the appropriate (continuous) Gaussian distribution function, as the mean evolves from 1 to 50.

As the mean becomes large, the Poisson distribution can be approximated by a normal distribution with the same mean and standard deviation as the Poisson distribution it is approximating. Recall that for a Poisson distribution, we have that the standard deviation is defined in terms of the mean, $\sigma = \sqrt{N}$. Thus from eqns (3.11) and (3.7) we can write the approximation as:

$$\frac{\exp\left(-\overline{N}\right)\overline{N}^{N}}{N!} \simeq \frac{1}{\sigma\sqrt{2\pi}}\exp\left[-\frac{(x-\overline{x})^{2}}{2\sigma^{2}}\right] \simeq \frac{1}{\sqrt{2\pi\overline{x}}}\exp\left[-\frac{(x-\overline{x})^{2}}{2\overline{x}}\right],$$

where we use the continuous variable x for the Gaussian curve.

From Fig. 3.6 it is clear that as the mean increases the Poisson distribution becomes more symmetric and the approximation becomes increasingly good. As a rule of thumb, once $\overline{N} \geq 35$ the asymmetry in the Poisson distribution is negligible, and it can be approximated to a normal distribution.

3.5   The central limit theorem

The **central limit theorem** (CLT) is a theorem from statistics of great importance; see, for example Squires (2001, Section 3.8) and Lyons (1991, Section 1.11.3). One way of stating the theorem is as follows: the sum of a large number of independent random variables, each with finite mean and variance, will tend to be normally distributed, irrespective of the distribution function of the random variable.

Note that:

(1) Peculiarly, a normal (Gaussian) distribution is obtained for any distribution of the individual measurements.[3]

(2) Equation (2.1), which defines the average, is a sum of a number of independent random variables. Hence the CLT applies to the statistics of the evaluation of the mean.

(3) The resulting normal distribution will have the same mean as the parent distribution, but a smaller variance. In fact, the variance is equal to the variance of the parent divided by the sample size. This is the mathematical statement which underpins our observation in Section 2.7 of the improvement in precision of estimating the mean with increasing signal-to-noise, and of the reduction of the standard error (standard deviation of the mean) by a factor of $\sqrt{N}$ with respect to the standard deviation of the sample when N data points are collected.

(4) The agreement between the distribution of the sum and a Gaussian only becomes exact in the (unphysical) limit of an infinite number of measurements. Fortunately, for most 'reasonable' distribution functions (which experimental measurements tend to follow), the agreement is very good for a small number of measurements.

[3] 'Any' should be interpreted here as meaning any kind of function which is likely to occur as a distribution function for experimental measurements; mathematicians will be able to think of more exotic functions for which the central limit theorem will not hold.

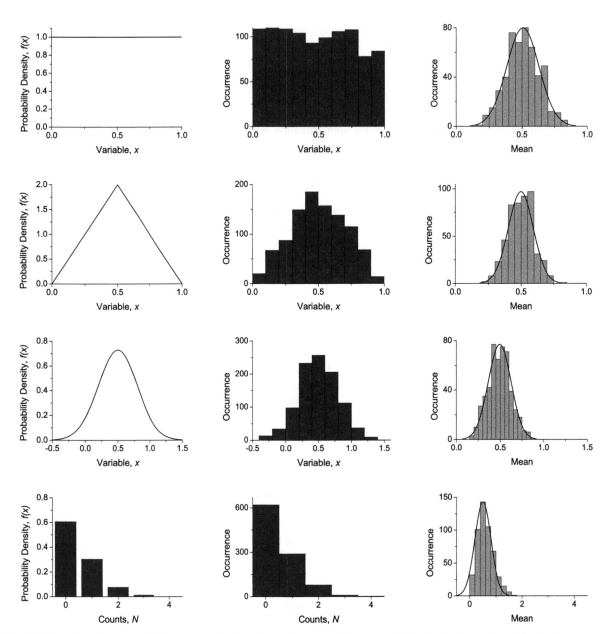

Fig. 3.7 Illustration of the central limit theorem. The first column shows four different probability distribution functions. The first three are for the continuous variable x, and have a mean of $\bar{x} = 0.5$. The last is a Poisson distribution with mean count $\overline{N} = 0.5$. The second column plots the outcome of 1000 computer-generated trials of choosing the variable from the appropriate distribution function. There are statistical fluctuations in the computer-generated data, but the shapes of the sample distributions match those of the population distribution. The third column shows the results of another computer-generated set of experiments, where five numbers are chosen from the parent distribution and averaged; this procedure was repeated 1000 times. The distribution of these 1000 means is plotted, and is seen in each case to follow a Gaussian distribution (the continuous curve superimposed on the histograms). The standard deviation of the distributions in the third column is $\sqrt{5}$ smaller than the parent standard deviations in the first columns as five data points were used to generate the average.

(5) We will use the central limit theorem as justification for assuming that the distribution function of interest to us is always a Gaussian (with the exception of Poisson counting statistics), and apply quantitative results, such as confidence limits, valid for Gaussian distributions.

More mathematically, the CLT states that if $x_1, x_2, \ldots, x_N$ are N independent random variables drawn from any distribution which has a mean $\bar{x}$ and standard deviation σ, then the distribution of the sample mean, which is $\frac{1}{N}(x_1 + x_2 + \cdots + x_N)$, is normal with a mean $\bar{x}$ and standard deviation $\frac{\sigma}{\sqrt{N}}$.

3.5.1 Examples of the central limit theorem

We now give three different examples of the central limit theorem, one of which uses data generated on a computer; the other two involve actual measurements.

Figure 3.7 shows four different probability density functions in the first column. The first three are for a continuous variable x, and each has a mean of $\bar{x} = 0.5$. The first distribution is the uniform distribution, the second a triangular distribution (both of the above have the range $0 \leq x \leq 1$). The third is a Gaussian distribution of mean $\bar{x} = 0.5$ and standard deviation $\sigma = 0.3$. The fourth is a discrete Poisson distribution, with a mean count of $\overline{N} = 0.5$. The second column shows the results of trials where 1000 points are chosen according to the relevant probability distribution function. Unsurprisingly, these figures have statistical noise, but are seen to have the same shape as the corresponding mathematical function. The third column depicts the result of 1 000 trials, each of which chose five samples from the appropriate distribution and from which the mean was calculated. The distribution of the means is shown here. There are three things to note: (i) irrespective of the shape of the initial probability distribution function, the histogram of the means is well described by a normal distribution; (ii) the distribution of means is peaked at 0.5, the mean of the original distributions; (iii) the width of the distribution of means is less than the width of the original probability distribution functions, by a factor $\sqrt{N}$. This is a manifestation of the reduction of the standard deviation of the means by $\sqrt{5}$ with respect to the standard deviation of the original probability distribution functions, in agreement with the discussion of the standard error in Chapter 2. These three outcomes are in agreement with the predictions of the central limit theorem.

An example of a situation where the distribution of measurements is non-Gaussian is radioactive decay, for which a Poisson distribution is obtained. Figure 3.8(a) shows a typical histogram obtained after making 58 measurements of 1 s duration. The mean number of counts per second is 7.2, and the histogram shows the characteristic asymmetric profile of a Poisson distribution. As an illustration of the central limit theorem the experiment was repeated another 50 times. Figure 3.8(b) shows the distribution of the means of the 51 trials,

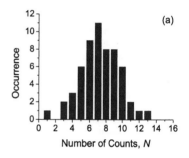

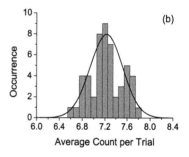

Fig. 3.8 Part (a) shows the result of a radioactive decay experiment. 423 counts were recorded in 58 seconds; the histogram shows the occurrences of the number of counts in one-second intervals about a mean of 7.3 counts per second. On repeating this experiment 51 times, it is possible to plot the distribution of the means, as is done in (b). Here it is seen that the distribution of the means is (i) well described by a Gaussian, and (ii) significantly narrower than the sample Poisson distribution.

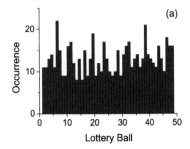

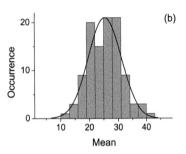

Fig. 3.9 Part (a) shows a histogram of the occurrences of the 49 balls in all 106 national lottery draws for the year 2000. If there was no bias a uniform distribution of 13 occurrences per ball would be expected; the experimental data are in good agreement with this within the statistical fluctuations. Six balls are chosen in each draw, the average number was calculated, and the histogram of the 106 results is plotted. As expected, a Gaussian distribution is obtained for the means, with a narrower standard deviation (by a factor of $\sqrt{6}$) compared with the parent distribution.

together with the best-fit Gaussian distribution. It is evident that the distribution of the means is symmetric, unlike the distribution of individual counts. A discussion of how well the experimental results follow the theoretical model is delayed until Chapter 8. The last example we discuss here is depicted in Fig. 3.9. The national lottery in the UK has six balls chosen at random from a set numbered $1, \ldots 49$. The occurrence of any particular ball should be independent of the others, hence the relevant distribution function is the uniform one. Figure 3.9(a) shows the outcomes of all 106 draws in the year 2000. The distribution of individual balls is seen to be approximately uniform, centred on the mean of 13 occurrences per ball. Figure 3.9(b) shows the average number of the six balls for the 106 draws; the average is obtained by summing the six integer ball numbers, and dividing by six. Once again, although the parent probability distribution function is uniform, the distribution of means is seen to follow a Gaussian distribution, in accordance with the central limit theorem. Further analysis of the mean and standard deviation of the distributions is detailed in Exercise (3.10).

Why is the distribution of means narrower than the distribution of the individual measurements? Consider the lottery example again. The smallest possible mean is 3.5, which is achieved uniquely from the numbers 1, 2, 3, 4, 5 and 6. In contrast, to obtain a mean of 25 we could have 22, 23, 24, 26, 27 and 28; or 1, 22, 25, 26, 27 and 49; or 1, 10, 25, 26, 39 and 49; or 3, 9, 14, 31, 45 and 48; or …. There are many more sets of six integers chosen at random from 1–49 with a mean of 25 than there are with a mean of 3.5. The argument holds for means much larger than the most likely one also: 44, 45, 46, 47, 48 and 49 is the only combination which yields the highest possible mean of 46.5. As each random sequence is as likely as any other one, it is far more likely statistically to obtain six numbers with a mean around 25 than 3.5. Consequently, the width of the distributions of means is significantly narrower than the distribution of individual numbers.

Chapter summary

- For a continuous probability distribution function $P_{DF}(x)$, the probability, P, that x lies between two values x_1 and x_2 is given by the area under the curve: $P(x_1 \leq x \leq x_2) = \int_{x_1}^{x_2} P_{DF}(x)\, dx$.
- The variance, σ^2, is the mean of the square minus the square of the mean: $\sigma^2 = \overline{x^2} - \bar{x}^2$.
- The standard deviation is the square root of the variance.
- The Gaussian, or normal, distribution function is specified by its centre and standard deviation: $G(x; \bar{x}, \sigma) = \dfrac{1}{\sigma\sqrt{2\pi}} \exp\left[-\dfrac{(x - \bar{x})^2}{2\sigma^2}\right]$.
- For a Gaussian distribution 68% of the data are expected to lie within the standard deviation of the mean; 95% of the data are expected to lie within two standard deviations of the mean; 99.7% of the data are expected to lie within three standard deviations of the mean.

- The distribution of discrete counts, e.g. radioactive decays, follow a Poisson distribution $P\left(N; \overline{N}\right) = \dfrac{\exp\left(-\overline{N}\right)\overline{N}^{N}}{N!}$.
- Only the mean, $\overline{N}$, is needed to specify a Poisson distribution.
- If N counts are detected in a given time, the error on this is $\sqrt{N}$.

Exercises

(3.1) *Discrete or continuous*
Which of the following variables are discrete, and which continuous? (i) The number of marks awarded for an examination paper; (ii) the height of adult males; (iii) the concentration of CO_2 in the atmosphere; (iv) the charge stored in a capacitor; and (v) the monthly salary of university employees.

(3.2) *Uniform distribution*
A probability distribution function of interest in error analysis is the uniform distribution. It is defined as

$$P_U\left(x; \bar{x}, a\right) = \begin{cases} 1/a & \text{if } \bar{x} - a/2 \leq x \leq \bar{x} + a/2, \\ 0 & \text{otherwise.} \end{cases}$$

Here the parameter $\bar{x}$ is the mean of the distribution, and a is the interval in which the probability distribution is uniform. Show that (i) the distribution $P_U\left(x; \bar{x}, a\right)$ is normalised; (ii) the mean of the distribution is indeed $\bar{x}$; (iii) the standard deviation is given by $\sigma = \dfrac{a}{\sqrt{12}}$.

(3.3) *Normal distributions*
Consult a reference resource and list three examples of naturally occurring distributions which are known to follow a Gaussian distribution.

(3.4) *Confidence limits for a Gaussian distribution*
Verify the results of Table 3.1 for the fraction of the data which lies within different ranges of a Gaussian probability distribution function. What fraction of the data lies outside the following ranges from the mean? (i) $\pm 4\sigma$ and (ii) $\pm 5\sigma$. What is the (symmetric) range within which the following fractions of the data lie? (i) 50% and (ii) 99.9%.

(3.5) *Calculations based on a Gaussian distribution*
Bags of pasta are sold with a nominal weight of 500 g. In fact, the distribution of weight of the bags is normal with a mean of 502 g and a standard deviation of 14 g. What is the probability that a bag contains less than 500 g? In a sample of 1000 bags how many will contain at least 530 g?

(3.6) *Identifying a potential outlier*
Seven successive measurements of the charge stored on a capacitor (all in μC) are: 45.7, 53.2, 48.4, 45.1, 51.4, 62.1 and 49.3. The sixth reading appears anomalously large. Apply Chauvenet's criterion to ascertain whether this data point should be rejected. Having decided whether to keep six or seven data points, calculate the mean, standard deviation and error of the charge.

(3.7) *Calculations based on a Poisson distribution (1)*
In the study of radioactive decay 58 successive experiments for one second yielded the following counts (these are the data plotted in Fig. 3.8).

N	1	3	4	5	6	7
Occurrence	1	2	3	6	9	11
N	8	9	10	11	12	13
Occurrence	8	8	6	2	1	1

Calculate (i) the total number of counts recorded; (ii) the mean count; and (iii) the mean count rate. Assuming that the data are well described by a Poisson distribution and that another 58 one-second counts are recorded, calculate (i) the expected number of occurrences of five counts or fewer; (ii) the expected number of occurrences of 20 counts or more.

(3.8) *Calculations based on a Poisson distribution (2)*
In the study of radioactive decay during a one-minute period 270 counts are recorded. Calculate: (i) the mean count rate; (ii) the error in the mean count rate; and (iii) the fractional error in the mean count rate. Were the experiment to be repeated with a 15-minute counting interval, what is (iv) the expected count; and (v) the probability of obtaining exactly 270×15 counts?

(3.9) *Approximation for high means*

Plot a histogram of a Poisson distribution with mean 35. Using the same axes plot the continuous function of a Gaussian with a mean of 35, and standard deviation $\sqrt{35}$. Comment on similarities and differences between the distributions.

(3.10) *An example of the central limit theorem*

The lottery results encapsulated in Fig. 3.9 are based on six numbers being selected from the integers $1, 2, \ldots, 49$. The distribution of these numbers should follow the functional form given in Exercise (3.2). Use the results of that exercise to predict the mean and standard deviation expected for this distribution. How do these compare with the results for the year 2000, with a mean of 25.4, and standard deviation 14.3? The lottery has six numbers selected, with the mean readily calculated. Based on the central limit theorem we expect the distribution of the means to follow a Gaussian distribution with a standard deviation which is $\sqrt{6}$ smaller than the original distribution. What do you predict for the mean and standard deviation of the means? How do these compare with the results for the year 2000, with a mean of 25.4, and standard deviation 5.7?

Error propagation

4

The aim of most experiments in the physical sciences is to combine several variables into a single quantity. The error on the combined value is a function of the errors on the constituent terms. As the addition of probabilities is not linear, simply summing the errors of the constituent terms gives an overestimate of the error in the combined variable. In this chapter we show how errors can be propagated through single and multi-variable functions using a functional approach (highly amenable to spreadsheet analysis) and a calculus-based approximation. We provide look-up tables for commonly encountered functions and discuss experimental strategy based on the dominant error.

4.1 Propagating the error in a single-variable function

If one measures a variable A to have a mean $\bar{A}$ and standard error α_A, it is instructive to see how it propagates through a single-variable function $Z = f(A)$. The best estimate of Z will be $\bar{Z} = f(\bar{A})$. In contrast, the uncertainty in Z is a function of *both* A and its uncertainty.

In Fig. 4.1 we show Bragg's law which relates the X-ray wavelength, λ, to the incident Bragg angle, θ. In this example, $\lambda = f(\theta)$, or more explicitly:

$$\lambda = 2d \sin \theta. \tag{4.1}$$

For a given angle, $\bar{\theta}$, one can calculate the wavelength through eqn (4.1). However, as eqn (4.1) contains a nonlinear relationship, the uncertainty in wavelength, α_λ, depends both on the angle, θ, and its uncertainty, α_θ. Note that in Fig. 4.1(b) a symmetric error in the angle, α_θ, maps into an asymmetric error in wavelength.

If α_A represents the error on the mean $\bar{A}$, the error in the function Z, α_Z, can be found by propagating $\bar{A} \pm \alpha_A$ through the function. Thus:

$$\bar{Z} \pm \alpha_Z = f(\bar{A} + \alpha_A), \tag{4.2}$$

$$\bar{Z} = f(\bar{A}), \tag{4.3}$$

$$\bar{Z} \mp \alpha_Z = f(\bar{A} - \alpha_A). \tag{4.4}$$

The origin of the $\pm$ signs in eqn (4.2) and (4.4) is as follows: $f(\bar{A} + \alpha_A)$ will give $\bar{Z} + \alpha_Z$ if the gradient of $f(A)$ with respect to Z is positive, and $\bar{Z} - \alpha_Z$

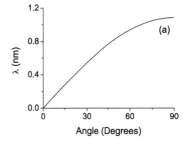

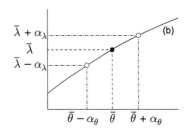

Fig. 4.1 (a) A plot of Bragg's law relating the X-ray wavelength, λ, to the incident angle, θ, for a Si single crystal. (b) Enlarged section showing how an uncertainty in the incident angle, α_θ, maps directly into an uncertainty in wavelength, α_λ. As this is a nonlinear relationship α_λ depends both on θ and α_θ.

if the gradient of $f(A)$ is negative. The error propagation using eqns (4.2) and (4.4) is shown schematically for Bragg's law in Fig. 4.1(b). As the gradient of $f(\theta)$ with respect to λ is positive, $f(\bar{\theta} + \alpha_\theta)$ returns the positive error in λ.

4.1.1 The functional approach for single-variable functions

Spreadsheets greatly facilitate the propagation of errors in a single-variable function. The error (assumed here to be symmetric) in Z for a particular measurement $\bar{A}$ is simply:

$$\alpha_Z = \left| f(\bar{A} + \alpha_A) - f(\bar{A}) \right|. \tag{4.5}$$

This relationship is valid for any single-variable function. Many commonly encountered data-analysis packages include built-in functions such as standard trigonometric functions, logarithms, exponentials, powers and even special functions such as Bessel and gamma. The argument of the function can easily be calculated at $\bar{A}$ and $\bar{A} \pm \alpha_A$, enabling a quick determination of the error in Z using eqn (4.5).

Returning to the Bragg's law example, eqn (4.5) can be used to calculate the error in the wavelength as a function of the incident angle; this is shown in Fig. 4.2. An uncertainty of $\alpha_\theta = 0.050°$ will give an error of $\alpha_\lambda = 0.92$ pm at $\theta = 15°$ and $\alpha_\lambda = 0.25$ pm at $\theta = 75°$. The error as a function of incident angle is, in fact, a cosine function for reasons that will become clear in Section 4.1.2.

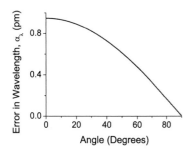

Fig. 4.2 The error in the wavelength, α_λ, as a function of incident angle, θ, using Bragg's law. The error in the incident angle, α_θ, is a constant $0.050°$. The error in the wavelength has a cosine dependence on the angle.

4.1.2 A calculus-based approximation for single-variable functions

We have seen in Fig. 4.1 how the error could be propagated through a single-variable function. An approach based on calculus is highlighted in Fig. 4.3 where a nonlinear function, $Z = f(A)$, and its tangent at point $P = (\bar{A}, f(\bar{A}))$ are sketched. The gradient of the tangent can be found using the triangle PQR where the gradient of the tangent is $\dfrac{QR}{PQ}$.

In the limit that α_A is small we can approximate the coordinates of R and S to be equal. (This is equivalent to retaining only the first two terms of the Taylor series expansion of eqn (4.5) and is discussed further in Section 4.2.3.) Mathematically, using the coordinates from Fig. 4.3(b), we can write:

$$f(\bar{A}) + \frac{\mathrm{d}f(A)}{\mathrm{d}A} \alpha_A = f(\bar{A} + \alpha_A), \tag{4.6}$$

which, using eqn (4.5) and recalling that $\dfrac{\mathrm{d}Z}{\mathrm{d}A} \equiv \dfrac{\mathrm{d}f(A)}{\mathrm{d}A}$, leads to the general result for a single-variable function:

$$\alpha_Z = \left| \frac{\mathrm{d}Z}{\mathrm{d}A} \right| \alpha_A. \tag{4.7}$$

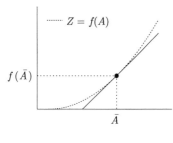

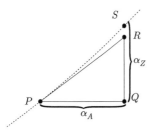

Fig. 4.3 The upper part shows a nonlinear function $Z = f(A)$ and its tangent at $\bar{A}$. The lower part shows an expanded view in the vicinity of $\bar{A}$: the four points P, Q, R and S have co-ordinates $(\bar{A}, f(\bar{A}))$, $(\bar{A} + \alpha_A, f(\bar{A}))$, $\left(\bar{A} + \alpha_A, f(\bar{A}) + \frac{\mathrm{d}f}{\mathrm{d}A} \times \alpha_A\right)$ and $(\bar{A} + \alpha_A, f(\bar{A} + \alpha_A))$, respectively.

There are three things to note about eqn (4.7). Firstly, the modulus sign for the gradient ensures that the error in Z is a positive number. Secondly, the calculus method predicts symmetric error bars—in contrast the functional approach of eqns (4.2) and (4.4) allows for asymmetric error bars. Thirdly, this result is only valid for small errors.

The calculus approximation requires the ability to calculate the derivative of the function at specific arguments. For complicated functions, these calculations can become non-trivial. There are certain functions, for example $Z(\theta) = \sin^{-4}(\theta/2)$, for which the spreadsheet approach is considerably simpler than the calculus approach.[1]

[1] The function $\sin^{-4}(\theta/2)$ describes the angular dependence of the differential cross-section in Rutherford scattering.

4.1.3   Look-up table for common single-variable functions

We can use the calculus method, eqn (4.7), to derive a look-up table to help propagate errors through commonly encountered single-variable functions; see Table 4.1.

Table 4.1 Results for the propagation of errors in single-variable functions. The results for the trigonometric functions assume that the angles and their errors are in radians.

Function, $Z(A)$	$\dfrac{dZ}{dA}$	Error
$\dfrac{1}{A}$	$-\dfrac{1}{A^2}$	$\alpha_Z = \dfrac{\alpha_A}{A^2} = Z^2 \alpha_A$ OR $\left\vert\dfrac{\alpha_Z}{Z}\right\vert = \left\vert\dfrac{\alpha_A}{A}\right\vert$
$\exp A$	$\exp A$	$\alpha_Z = \exp A \, \alpha_A = Z \, \alpha_A$
$\ln A$	$\dfrac{1}{A}$	$\alpha_Z = \dfrac{\alpha_A}{A}$
$\log A$	$\dfrac{1}{\ln(10) \, A}$	$\alpha_Z = \dfrac{\alpha_A}{\ln(10) \, A}$
A^n	$n A^{n-1}$	$\alpha_Z = \left\vert n A^{n-1}\right\vert \alpha_A$ OR $\left\vert\dfrac{\alpha_Z}{Z}\right\vert = \left\vert n\dfrac{\alpha_A}{A}\right\vert$
10^A	$10^A \ln(10)$	$\alpha_Z = 10^A \ln(10) \, \alpha_A$
$\sin A$	$\cos A$	$\alpha_Z = \vert\cos A\vert \, \alpha_A$
$\cos A$	$-\sin A$	$\alpha_Z = \vert\sin A\vert \, \alpha_A$
$\tan A$	$1 + \tan^2 A$	$\alpha_Z = \left(1 + Z^2\right) \alpha_A$

4.1.4   Worked example—single variable function

We will illustrate both approaches to the propagation of errors through the single-variable function $Z = 10^A$. Suppose we had measured $A = 2.3 \pm 0.1$. What is the value of Z and its error?

Our best estimate of Z is the mean, $\bar{Z} = 10^{2.3} = 199.5$. We defer the rounding of the mean until the error has been calculated.

(a) Calculating the error in Z using the functional approach:
The errors in Z are:

$$\alpha_Z^+ = 10^{2.3+0.1} - 10^{2.3} = 51.7,$$

and

$$\alpha_Z^- = 10^{2.3} - 10^{2.3-0.1} = 41.0.$$

We therefore say that our best estimate of Z lies within the range $158 \le \bar{Z} \le 251$. An approximation to the *symmetric* error bar can be found by quoting the average error bar. As we only know the uncertainty in A to one significant figure, we report our best estimate of Z as $Z = (2.0 \pm 0.5) \times 10^2$. An alternative way to report this result, which keeps the assymetry of the error bar, is $Z = \left(2.0^{+0.5}_{-0.4}\right) \times 10^2$.

(b) Calculating the error in Z using the calculus approximation and the look-up tables:

$$\alpha_Z = Z \ln(10)\, \alpha_A = 199.5 \times \ln(10) \times 0.1 = 45.9,$$

again, as we only know the uncertainty in A to one significant figure, we quote

$$Z = (2.0 \pm 0.5) \times 10^2.$$

We note that it is only the functional approach that shows the asymmetry in mapping the errors in A to those in Z. As the error is relatively small, the calculus method is a good approximation and the two methods yield identical results.

For larger errors in A the calculus approximation becomes less reliable. Suppose we had measured $A = 2.3 \pm 0.4$ instead. Our best estimate of Z remains unchanged. Calculating the errors using the functional approach shows Z now lies within the range $79 \le \bar{Z} \le 501$. We illustrate the asymmetry of the error bar[2] by reporting our result as $Z = \left(2^{+3}_{-1}\right) \times 10^2$. In contrast, the calculus-based approach yields $Z = (2 \pm 2) \times 10^2$.

[2] The asymmetry of the error bars determined using the functional approach indicate that the distribution of Z is not Gaussian. We therefore do not know trivially the confidence limits to which these error bars correspond, and cannot blindly apply the results from Chapter 3.

4.2   Propagating the error through a multi-variable function

In many cases the function through which we wish to propagate our errors is multi-variable. One needs to map a series of measurements, $A, B, C, \ldots$ and their associated errors through the function $Z = f(A, B, C, \ldots)$. As in the case for a single-variable function, our best estimate of Z is made through the mean values of the parameters $\bar{Z} = f\left(\bar{A}, \bar{B}, \bar{C}, \ldots\right)$. As before, the error in Z is a function of both the mean values and their errors: $\alpha_Z = f\left(\bar{A}, \bar{B}, \bar{C}, \ldots; \alpha_A, \alpha_B, \alpha_C, \ldots\right)$.

4.2.1 The functional approach for multi-variable functions

Consider first a function of two variables, $Z = f(A, B)$. The equivalent of Fig. 4.1 is now a two-dimensional surface, as shown in Fig. 4.4 (a). Our best estimate of Z is a point on the surface given by $\bar{Z} = f(\bar{A}, \bar{B})$. The error in Z comprises two components. One component is the change in Z (the height of the surface) around the mean when A is varied and B is kept constant, as seen in Fig. 4.4 (b). This change of height is:

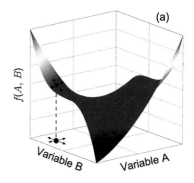

$$\alpha_Z^A = f(\bar{A} + \alpha_A, \bar{B}) - f(\bar{A}, \bar{B}). \tag{4.8}$$

Similarly, as shown in Fig. 4.4(c), there is also a change in the surface around the mean when B is changed and A remains fixed:

$$\alpha_Z^B = f(\bar{A}, \bar{B} + \alpha_B) - f(\bar{A}, \bar{B}). \tag{4.9}$$

To proceed, we need to assume that the uncertainties in A and B are **uncorrelated** and that A and B are **independent variables**. (An independent variable is not correlated with either the magnitude or error of any other parameter.) The total error in Z is obtained by applying Pythagoras' theorem and adding the components in quadrature. For N independent variables we apply Pythagoras' theorem in N dimensions to obtain the general result:

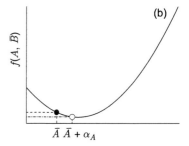

$$(\alpha_Z)^2 = \left(\alpha_Z^A\right)^2 + \left(\alpha_Z^B\right)^2 + \left(\alpha_Z^C\right)^2 + \cdots \tag{4.10}$$

We see in Fig. 4.4 that eqn (4.10) can be written in functional form as:

$$(\alpha_Z)^2 = \left[f(\bar{A} + \alpha_A, \bar{B}, \bar{C}, \ldots) - f(\bar{A}, \bar{B}, \bar{C}, \ldots)\right]^2$$
$$+ \left[f(\bar{A}, \bar{B} + \alpha_B, \bar{C}, \ldots) - f(\bar{A}, \bar{B}, \bar{C}, \ldots)\right]^2$$
$$+ \left[f(\bar{A}, \bar{B}, \bar{C} + \alpha_C, \ldots) - f(\bar{A}, \bar{B}, \bar{C}, \ldots)\right]^2$$
$$+ \cdots \tag{4.11}$$

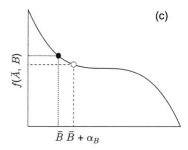

As we discussed for the single-variable case, this approach is very amenable to spreadsheet analysis as the error in Z is found by quickly re-evaluating the function with different arguments. It is straightforward to see how eqn (4.2) and eqn (4.4) can be used to look for asymmetries in the error bars when propagating an error in the single-variable case. However, in N-dimensions this becomes difficult, and eqn (4.11) assumes that the magnitude of the variation in the function Z is the same when an independent variable is either increased or decreased by an error bar around its mean position. This approximation usually holds for small errors but one should be careful if the function is highly nonlinear, or the error bars are large. In such cases, it is wise to look more carefully at the error surface, a concept which is discussed in more detail in Chapter 6.

Fig. 4.4 (a) A two-dimensional surface plot of the function $Z = f(A, B)$. (b) A slice along the A-axis when B is kept at its mean value $\bar{B}$. The change in height of the surface, α_Z^A, owing to a displacement α_A along the A-axis from the point $\bar{Z} = f(\bar{A}, \bar{B})$ is $f(\bar{A} + \alpha_A, \bar{B}) - f(\bar{A}, \bar{B})$. (c) A slice along the B-axis when A is kept at its mean value $\bar{A}$. The change in height of the surface, α_Z^B, owing to a displacement α_B along the B-axis from the point $\bar{Z} = f(\bar{A}, \bar{B})$ is $f(\bar{A}, \bar{B} + \alpha_B) - f(\bar{A}, \bar{B})$. Pythagoras' theorem relates the total error, α_Z, to the components: $(\alpha_Z)^2 = \left(\alpha_Z^A\right)^2 + \left(\alpha_Z^B\right)^2$.

4.2.2   Worked example—functional approach for multi-variable functions

[3] The van der Waals equation of state is a correction to the ideal gas law. The coefficients a and b take into account, respectively, the attraction between the constituents of the gas, and the volume excluded owing to the finite size of the atoms or molecules. Both of these terms are assumed to be zero in the case of an ideal gas. These specific corrections to the ideal gas law were proposed by Johannes van der Waals in 1873.

The van der Waals equation of state for a gas is[3]

$$\left(P + \frac{a}{V_m^2}\right)(V_m - b) = RT, \tag{4.12}$$

where P is the pressure, V_m is the molar volume, T is the absolute temperature, R is the universal gas constant, with a and b being species-specific van der Waals coefficients. Calculate (i) the pressure of a sample of nitrogen with molar volume $V_m = (2.000 \pm 0.003) \times 10^{-4}\,\mathrm{m^3\,mol^{-1}}$ at a temperature of $T = 298.0 \pm 0.2\,\mathrm{K}$, given the van der Waals coefficients for nitrogen are $a = 1.408 \times 10^{-1}\,\mathrm{m^6\,mol^{-2}\,Pa}$, and $b = 3.913 \times 10^{-5}\,\mathrm{m^3\,mol^{-1}}$; (ii) the uncertainty in the pressure. Take R to be $8.3145\mathrm{J\,K^{-1}\,mol^{-1}}$.

(i) Rearranging eqn (4.12) to make P the subject gives

$$P(V_m, T) = \frac{RT}{V_m - b} - \frac{a}{V_m^2}. \tag{4.13}$$

Inserting the numbers from the question, and recalling that $1\,\mathrm{J} \equiv 1\,\mathrm{m^3\,Pa}$, we obtain the best estimate for the pressure of the gas

$$P\left(\overline{V_m}, \overline{T}\right) = 11.882\,\mathrm{MPa}.$$

Note we keep five significant figures at this stage; the number of significant figures we report for the mean is ascertained after the error has been calculated.

(ii) The uncertainties in both temperature and molar volume contribute to the uncertainty in the pressure. To use eqn (4.11) we need to evaluate $P\left(\overline{V_m} + \alpha_V, \overline{T}\right)$ and $P\left(\overline{V_m}, \overline{T} + \alpha_T\right)$. Explicitly, the expressions are:

$$P\left(\overline{V_m} + \alpha_V, \overline{T}\right) = \frac{R\overline{T}}{\overline{V_m} + \alpha_V - b} - \frac{a}{\left(\overline{V_m} + \alpha_V\right)^2} = 11.864\,\mathrm{MPa},$$

and

$$P\left(\overline{V_m}, \overline{T} + \alpha_T\right) = \frac{R\left(\overline{T} + \alpha_T\right)}{\overline{V_m} - b} - \frac{a}{\overline{V_m}^2} = 11.892\,\mathrm{MPa}.$$

The contribution to the error in the pressure due to the temperature is $\alpha_P^T = P\left(\overline{V_m}, \overline{T} + \alpha_T\right) - P\left(\overline{V_m}, \overline{T}\right)$, and similarly for the contribution to the error due to the volume, α_P^V. These contributions are evaluated to be

$$\alpha_P^T = 0.010\,\mathrm{MPa}, \qquad \alpha_P^V = 0.018\,\mathrm{MPa}.$$

As both contributions are similar, eqn (4.10) is used to find the uncertainty in P:

$$\alpha_P = \sqrt{\left(\alpha_P^T\right)^2 + \left(\alpha_P^V\right)^2} = 0.021\,\mathrm{MPa}.$$

As the uncertainties in T and V_m were quoted to one significant figure, we quote the uncertainty in P to one significant figure: $\alpha_P = 0.02\,\text{MPa}$. Finally, we decide how many decimal places to retain for the mean, and quote the result as $P = 11.88 \pm 0.02\,\text{MPa}$.

4.2.3   A calculus approximation for multi-variable functions

A calculus-based approximation for the propagation of errors through multi-variable functions can be derived by Taylor series expansion[4] of the terms in square brackets in eqn (4.11). This procedure yields, for each term in eqn (4.11), an expression similar to:

$$f\left(\bar{A} + \alpha_A, \bar{B}, \bar{C}, \ldots\right) = f\left(\bar{A}, \bar{B}, \bar{C}, \ldots\right) + \left.\frac{\partial f}{\partial A}\right|_{A=\bar{A}} \times \alpha_A$$

$$+ \frac{1}{2}\left.\frac{\partial^2 f}{\partial A^2}\right|_{A=\bar{A}} \times \alpha_A^2 + \cdots \qquad (4.14)$$

Equation (4.14) contains the function evaluated at the mean, and subsequent terms proportional to the partial derivatives of the function. An implicit assumption of the calculus-based approximation is that the magnitude of the error is small. Thus, second and higher order derivate terms are negligible compared to the gradient term and consequently are not included. In the derivation of eqn (4.11) we assumed that the variables were independent, thus cross-terms involving products of uncertainties in two variables, $\dfrac{\partial^2 f}{\partial A\,\partial B} \times \alpha_A\alpha_B$, average to zero.[5]

Equation (4.14) thus gives the error in the function Z due to deviations in the variable A:

$$f\left(\bar{A} + \alpha_A, \bar{B}, \bar{C}, \ldots\right) - f\left(\bar{A}, \bar{B}, \bar{C}, \ldots\right) = \left(\left.\frac{\partial f}{\partial A}\right|_{A=\bar{A}} \times \alpha_A\right). \qquad (4.15)$$

We recognise this as a combination of the results of eqns (4.5) and (4.7). Hence the generalisation of eqn (4.7) to obtain an expression for the error in a multi-variable function $Z = f(A, B, C, \ldots)$ using the calculus approximation is:

$$(\alpha_Z)^2 = \left(\frac{\partial Z}{\partial A}\right)^2 (\alpha_A)^2 + \left(\frac{\partial Z}{\partial B}\right)^2 (\alpha_B)^2 + \left(\frac{\partial Z}{\partial C}\right)^2 (\alpha_C)^2 + \cdots \qquad (4.16)$$

Equation (4.16) highlights the fact that the contributions to the total error from independent variables are summed in quadrature, as depicted in Fig. 4.5.

4.2.4   A look-up table for multi-variable functions

We can apply the general calculus-based approximation for some common functions to produce a look-up table; see Table 4.2.

[4]Taylor's theorem enables a function to be expanded in a power series in x in a given interval, and states that if $f(x)$ is a continuous, single-valued function of x with continuous derivatives $f'(x), f''(x), \ldots$ in a given interval, then $f(x) = f(a) + \frac{(x-a)}{1!}f'(a) + \frac{(x-a)^2}{2!}f''(x) + \cdots$.   An alternative form, as used in the text, may be obtained by changing x to $a + x$.

[5]The cross-terms will be important in constructing the covariance matrix in Chapter 7.

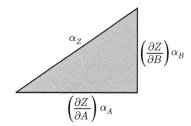

Fig. 4.5 As A and B are independent variables their contributions to the error in Z are orthogonal. The total error in Z is obtained by adding the contributions from A and B in quadrature, subject to Pythagoras' theorem.

Table 4.2 Some simple rules for the propagation of errors in multi-variable functions. *Always perform a quick check for dominant errors before using these formulae.*

Function, $Z(A)$	Expression used to calculate α_Z
$\left.\begin{array}{l} Z = A + B \\ Z = A - B \end{array}\right\}$	$\alpha_Z = \sqrt{(\alpha_A)^2 + (\alpha_B)^2}$
$\left.\begin{array}{l} Z = A \times B \\ Z = \dfrac{A}{B} \end{array}\right\}$	$\dfrac{\alpha_Z}{Z} = \sqrt{\left(\dfrac{\alpha_A}{A}\right)^2 + \left(\dfrac{\alpha_B}{B}\right)^2}$
$Z = A^n$	$\left\|\dfrac{\alpha_Z}{Z}\right\| = \left\|n\dfrac{\alpha_A}{A}\right\|$
$Z = kA$	$\alpha_Z = \|k\|\alpha_A \;\; \text{OR} \;\; \left\|\dfrac{\alpha_Z}{Z}\right\| = \left\|\dfrac{\alpha_A}{A}\right\|$
$Z = k\dfrac{A}{B}$	$\dfrac{\alpha_Z}{Z} = \sqrt{\left(\dfrac{\alpha_A}{A}\right)^2 + \left(\dfrac{\alpha_B}{B}\right)^2}$
$Z = k\dfrac{A^n}{B^m}$	$\dfrac{\alpha_Z}{Z} = \sqrt{\left(n\dfrac{\alpha_A}{A}\right)^2 + \left(m\dfrac{\alpha_B}{B}\right)^2}$
$Z = A + B - C + D$	$\alpha_Z = \sqrt{(\alpha_A)^2 + (\alpha_B)^2 + (\alpha_C)^2 + (\alpha_D)^2}$
$Z = \dfrac{(A \times B)}{(C \times D)}$	$\dfrac{\alpha_Z}{Z} = \sqrt{\left(\dfrac{\alpha_A}{A}\right)^2 + \left(\dfrac{\alpha_B}{B}\right)^2 + \left(\dfrac{\alpha_C}{C}\right)^2 + \left(\dfrac{\alpha_D}{D}\right)^2}$
$Z = \dfrac{(A^n \times B^m)}{(C^p \times D^q)}$	$\dfrac{\alpha_Z}{Z} = \sqrt{\left(n\dfrac{\alpha_A}{A}\right)^2 + \left(m\dfrac{\alpha_B}{B}\right)^2 + \left(p\dfrac{\alpha_C}{C}\right)^2 + \left(q\dfrac{\alpha_D}{D}\right)^2}$

4.2.5 Comparison of methods

We will illustrate the two approaches to the propagation of errors through the multi-variable function:[6]

$$Z = \frac{(A - B)}{(A + B)}. \tag{4.17}$$

[6] This is a commonly encountered function for the amplitude-reflection coefficient of a wave at a boundary between two media of impedance A and B.

Suppose we have measured $\bar{A} = 1000$ and $\bar{B} = 80$ both with 1% errors. Our best estimate of Z is $\bar{Z} = \dfrac{(\bar{A} - \bar{B})}{(\bar{A} + \bar{B})} = 0.852$.

(a) Calculating the error in Z using the calculus approximation: The error in Z is:

$$(\alpha_Z)^2 = \left(\frac{\partial Z}{\partial A} \cdot \alpha_A\right)^2 + \left(\frac{\partial Z}{\partial B} \cdot \alpha_B\right)^2$$

$$= \left(\frac{2B}{(A + B)^2} \cdot \alpha_A\right)^2 + \left(\frac{-2A}{(A + B)^2} \cdot \alpha_B\right)^2.$$

This gives, for the error in Z, $\alpha_Z = 0.00194$. As the errors in A and B are only quoted to one significant figure we report that $Z = 0.852 \pm 0.002$.

(b) Calculating the error in Z using the functional approach:

$$(\alpha_Z)^2 = \left[f\left(\bar{A} + \alpha_A, \bar{B}\right) - f\left(\bar{A}, \bar{B}\right) \right]^2 + \left[f\left(\bar{A}, \bar{B} + \alpha_B,\right) - f\left(\bar{A}, \bar{B}\right) \right]^2.$$

Using a spreadsheet to calculate the function at each argument:

$$f\left(\bar{A}, \bar{B}\right) = 0.8519, \; f\left(\bar{A} + \alpha_A, \bar{B}\right) = 0.8532, f\left(\bar{A}, \bar{B} + \alpha_B\right) = 0.8505$$

allows the error bar to be found, $\alpha_Z = 0.00193$. As the errors in A and B are only quoted to one significant figure we report that $Z = 0.852 \pm 0.002$.

The two methods give identical results. Note also that during any arithmetic calculation it is advisable to reduce the effects of rounding errors by maintaining a suitable number of significant figures and only rounding to the appropriate number of significant figures at the end.

4.2.6 Percentage errors—dominant error

Many of the expressions for the propagation of errors that arise in Table 4.1 are represented by the percentage error. It is frequently possible to bypass the need to use the rigorous expression if one performs a quick back of the envelope calculation. For example, if $Z = A \times B \times C \times D$, and A is known to 5%, and B, C and D to 1%, what is the percentage error in Z? The experienced practitioner will not have to use the appropriate formula from Table 4.2 as the addition of the percentage errors in quadrature will yield, to one significant figure, 5%.

RULE OF THUMB: Perform a quick calculation to identify any dominant errors. Consider whether a more rigorous calculation is useful.

4.2.7 Using the look-up tables

The look-up tables are ideal for propagating errors in multi-variable functions, particularly when summation or multiplication are involved. For example, the resonant frequency, f_0, in a circuit with an inductance, L, and capacitance, C, is:

$$f_0 = \frac{1}{2\pi} \frac{1}{\sqrt{LC}}. \tag{4.18}$$

The inductance and its error can be found by measuring the resonant frequency and the capacitance. The error in L can be found directly from the look-up tables:

$$L = \frac{1}{4\pi^2 f_0^2 C} \Rightarrow \left(\frac{\alpha_L}{L}\right)^2 = \left(2\frac{\alpha_{f_0}}{f_0}\right)^2 + \left(\frac{\alpha_C}{C}\right)^2 = 4\left(\frac{\alpha_{f_0}}{f_0}\right)^2 + \left(\frac{\alpha_C}{C}\right)^2.$$

$$\tag{4.19}$$

Note the factor of 4 in the contribution to the square of the percentage error in the inductance that arises on account of the inverse-squared dependence on the resonant frequency.

Occasionally the function of interest does not appear directly in the tables. Fortunately, the terms in the look-up table are applicable, after suitable modification, to many single-variable functions as well. For example:

$$Z(A, B) = kA + B, \tag{4.20}$$

where k is a constant. We can make the substitution $X(A) = kA$. Then the function $Z(A, B)$ becomes $Z = X(A) + B$. This functional form can be found in look-up Table 4.2, giving the error in Z as:

$$\alpha_Z = \sqrt{(\alpha_X)^2 + (\alpha_B)^2}. \tag{4.21}$$

We now need to calculate explicitly α_X, which is also in look-up Table 4.2:

$$X(A) = kA \Rightarrow \alpha_X = k\alpha_A. \tag{4.22}$$

Finally, we can substitute eqn (4.22) back into eqn (4.21) to find the expression for the error on Z:

$$\alpha_Z = \sqrt{k^2 (\alpha_A)^2 + (\alpha_B)^2}. \tag{4.23}$$

4.2.8   Using the look-up tables—health warning

It is important to ensure that the independence of the variables is maintained when making a substitution. The look-up tables are only valid for single-variable functions or variables. Recalling the example of eqn (4.17):

$$Z = \frac{(A - B)}{(A + B)}.$$

One might be tempted to use the look-up tables and make the substitutions $X(A, B) = (A - B)$ and $Y(A, B) = (A + B)$ such that $Z = X/Y$. If we then use the tables to compute the error it comes out to be $\alpha_Z = 0.012$, an order of magnitude larger than the correct value. This is because X and Y are now no longer independent variables but are correlated.

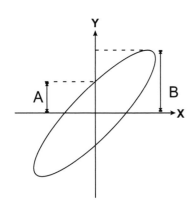

Fig. 4.6 The phase shift, δ, between two sine waves can be measured using the Lissajous method with an oscilloscope. The phase shift is the origin of the ellipticity of the ellipse in the figure, and is related to the quantities A and B via the equation $\delta = \arcsin\left(\frac{A}{B}\right)$.

Only substitute single-variable functions when using the look-up tables. Recall that the look-up tables are **only valid for independent variables** or single-variable functions.

4.3   Propagating errors in functions—a summary

In practice, error analysis is often an exercise that requires different approaches for different situations. We have detailed above the standard methods, but common sense should always be applied. To summarise the above:

- For simple functions which involve summation, multiplication and powers the look-up tables are the most convenient method.
- The calculus-based approach is only useful if the derivatives are easy to calculate.
- The functional approach, in which the function is calculated for various arguments, is particularly suitable for spreadsheet analysis.

Often a hybrid approach proves to be the most efficient method for determining the error in a multi-variable function. For example, consider the case depicted in Fig. 4.6 where the angle δ is defined as the arcsine of the ratio of A and B,

$$\delta = \arcsin\left(\frac{A}{B}\right). \tag{4.24}$$

In this case one could define a variable $C = A/B$. Using the look-up tables the error in C is quickly identified as:

$$\alpha_C = C\sqrt{\left(\frac{\alpha_A}{A}\right)^2 + \left(\frac{\alpha_B}{B}\right)^2}. \tag{4.25}$$

Given that C and its error have been determined, the easiest method to calculate the error in δ is through the functional approach. This method has the added advantage that, because arcsine is nonlinear, one could calculate the asymmetric error bars if the errors in C were large:

$$\alpha_\delta^\pm = \left|\arcsin\left(\bar{C}\right) - \arcsin\left(\bar{C} \pm \alpha_C\right)\right|. \tag{4.26}$$

Some of the issues regarding error propagation with a nonlinear function are highlighted in Fig. 4.7. Arcsine is a multi-valued function, i.e. there are many angles whose sine is 0.5. Care has to be exercised in deciding which branch of the function to choose. Arcsine is an example of a function whose argument has a restricted range; specifically $-1 \le C \le 1$. In Fig. 4.7(b) the argument is small, and eqn (4.26) can be used to calculate the assymetric error bars. In contrast, as depicted in Fig. 4.7(c), if the argument is large it will not be possible to use eqn (4.26) to calculate α_δ^+ when $\bar{C} + \alpha_C > 1$.

4.4   Experimental strategy based on error analysis

The uncertainties in A, B, C, etc. can have very different effects on the magnitude of the error in Z depending on its functional form. One should always con-

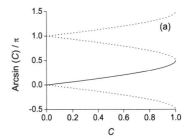

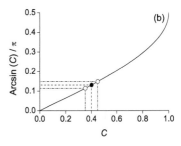

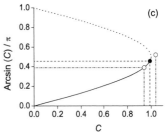

Fig. 4.7 (a) As arcsine is a multi-valued function, care has to be exercised in deciding which branch of the function to choose. In addition, the argument of arcsine has the restricted range $-1 \le C \le 1$. In (b) the argument is small, and eqn (4.26) is used to calculate the assymetric error bars. (c) When $\bar{C} + \alpha_C > 1$ it will not be possible to use eqn (4.26) to calculate α_δ^+.

centrate on reducing the **dominant error**, and not waste time trying to reduce the error on parameters which do not contribute significantly to the error on Z.

One example that is often encountered is when the functional form includes high-order powers. For example, the formula for the density, ρ, of a sphere of radius r and mass m is:

$$\rho = \frac{m}{V} = \frac{3m}{4\pi r^3}. \tag{4.27}$$

The different dependence on the parameters m and r of the density are shown in Fig. 4.8. Preliminary readings indicate that both the mass and radius were determined to a precision of 1%, and the radius to 1%. What should the strategy be to obtain the most precise measurement of the density?

From the look-up tables the fractional error in the density is:

$$\frac{\alpha_\rho}{\rho} = \sqrt{9\left(\frac{\alpha_r}{r}\right)^2 + \left(\frac{\alpha_m}{m}\right)} = \sqrt{9\left(\frac{1}{100}\right)^2 + \left(\frac{1}{100}\right)} = 3.2\%.$$

The error in the density is therefore dominated by the uncertainty in the radius—ignoring the error in the mass would give a 3% error in the density from the uncertainty in the measurement of r alone. The uncertainty in the mass hardly contributes to the error in the density (0.2%), therefore there is little point in measuring it more precisely. Only after the uncertainty in the radius has been decreased by at least one order of magnitude is it worth reconsidering the contribution from the error in the mass.

Another well-known case where particular care is needed is when the function of interest is the difference in two numbers of similar magnitude. Consider two parameters $A = 18 \pm 2$ and $B = 19 \pm 2$. The sum, $C = A + B = 37 \pm 3$, is known to a similar percentage error as either A or B. On the other hand, the difference $D = A - B = -1 \pm 3$ is poorly defined. Many experiments involve looking for small differences. One should always be aware that the fractional precision with which one can determine the difference is no longer trivially related to the precision of the individual parameters. A much better strategy is to find a physical parameter that is related directly to the difference. For example, consider the transmission of light through an optically active medium (e.g. sugar solution). One could measure the refractive index of the solution for right-hand circular polarisation, n_R, and then take a second measurement for left circular light, n_L. The degree of optical anisotropy is characterised by the difference, $n_R - n_L$, which is generally smaller than either n_R or n_L. Calculating the optical anisotropy using this method is prone to the large percentage error problem highlighted above. A better experimental technique is to illuminate the solution with plane polarised light. The polarisation axis rotates as the light propagates through the medium. Measuring the rotation angle is a direct measurement of $n_R - n_L$ and is typically recorded with a percentage error less than 1%.

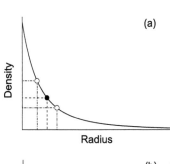

(a)

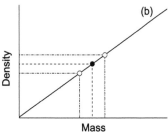

(b)

Fig. 4.8 The dependence of density on (a) radius and (b) mass. Owing to the inverse cubed dependence on radius the fractional error in density is three times larger than the fractional error in radius: $\frac{\alpha_\rho}{\rho} = 3\frac{\alpha_r}{r}$. By contrast, the linear dependence on mass means that the fractional uncertainties are equal: $\frac{\alpha_\rho}{\rho} = \frac{\alpha_m}{m}$.

Before beginning the definitive data set in an experiment identify the dominant source of error and concentrate your effort in reducing it. Pay particular attention when the calculation involves: (i) taking the difference between two nearly equal quantities; (ii) taking the power (>1) of a variable.

4.4.1 Experimental strategy for reducing the dominant error

The examples in the previous section highlighted the importance of identifying the dominant error. If one exists, there are two possible ways of reducing the uncertainty in the dominant error. (1) Persevere with the same apparatus: the standard deviation of the results is likely to be the same, but the uncertainty in the standard error on the mean will decrease (slowly) as the number of measurements is increased. (2) Find a better instrument, or method, to perform the experiment—this will result in the distribution of measurements having a smaller standard deviation, and hence a smaller standard error on the mean.

4.5 Combined experiments—the weighted mean

Suppose that there is a series of experiments, each of which measures the same parameter. The experiments could be attempting to calculate a particular variable in a number of different ways, or it could be the same experiment, but performed by different people. How do we combine all these separate measurements to yield the best value and its error? Before discussing how to combine different results, we first give this health warning.

When taking the weighted mean of a series of measurements, it is important that the compatibility of the results is considered: Combine multiple measurements of the same quantity only if they are consistent with each other. Consider outliers with care.

For example, if the result of an experiment to measure the speed of light is $c_1 = (3.00 \pm 0.01) \times 10^8$ m s^{-1} and another technique yields $c_2 = (4.00 \pm 0.02) \times 10^8$ m s^{-1} it is a waste of time to combine these results as the second value is obviously the subject of systematic errors.

Let the (consistent) results for one experiment be $x_i \pm \alpha_i$ and those of another be $x_j \pm \alpha_j$; see the graphs in Fig. 4.9. If the two results had errors of a similar magnitude the mean of the two readings would be the best estimate of the value, as this accords equal importance to the two experimental values:

$$\bar{x}_{i,j} = \frac{1}{2}\left(x_i + x_j\right). \tag{4.28}$$

It can be shown that the error in the above case is:

$$\frac{1}{\left(\alpha_{\bar{x}_{i,j}}\right)} = \sqrt{\frac{1}{\left(\alpha_i\right)^2} + \frac{1}{\left(\alpha_j\right)^2}}. \tag{4.29}$$

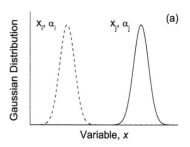

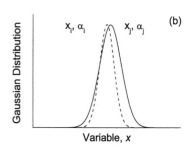

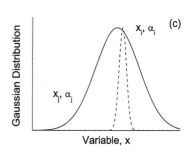

Fig. 4.9 Three circumstances which can occur when combining a pair of measurements. In (a) there is very little overlap between the distribution of results—there is no point trying to combine them. A better use of time would be trying to ascertain the origin of the discrepancy between the two sets of measurement. In (b) the two measurements can be combined, to yield a better estimate of the mean with a smaller uncertainty. In (c) the results are consistent, but there is little point combining them as adding the less precise measurement will hardly improve the more precise result.

In the more general case the error bars are different. Then, in the absence of any systematic error, we would naturally give more credence to the results with the smaller error. For a range of error values we treat each measurement with an importance, or weight, inversely proportional to the square of its standard error—in other words, the lower the standard error the more precise the measurement, and the greater its significance. One can derive an expression for this weighted mean and its uncertainty (Bevington and Robinson 2003, Chapter 4). The best combined estimate, x_{CE}, incorporating all of the available data is the sum of the weighted means, normalised by the sum of the weightings:

$$x_{CE} = \frac{\sum_i w_i x_i}{\sum_i w_i}, \tag{4.30}$$

where the weighting, w_i, is given by:

$$w_i = \frac{1}{\alpha_i^2}. \tag{4.31}$$

The inverse of the square of the standard error of the weighted mean is the sum of the weightings:

$$\frac{1}{\alpha_{CE}^2} = \frac{1}{\alpha_i^2} + \frac{1}{\alpha_j^2} + \frac{1}{\alpha_k^2} + \cdots = \sum_i \left(\frac{1}{\alpha_i^2}\right). \tag{4.32}$$

4.5.1   The error in the mean—a special case of the weighted mean

We can apply the formalism of the previous section to calculate the error in the mean. Consider N measurements of the same quantity, x_i, with $i = 1, \ldots, N$. We can think of each measurement as an estimate of the mean, and the best estimate will be their weighted sum. As there is two-thirds chance that each measurement will be within σ of the mean, we can take σ to be the error *in one of these measurements*. As the error is the same for each x_i, each measurement carries the same weight, and eqn (4.30) becomes:

$$x_{CE} = \frac{\sum_i w_i x_i}{\sum_i w_i} = \frac{\sum_i x_i}{N} = \overline{x}. \tag{4.33}$$

This gives the expected result that the combined estimate is the mean of the measurements. In a similar manner we can calculate the standard error of the weighted mean to be

$$\frac{1}{\alpha_{CE}^2} = \sum_i \left(\frac{1}{\sigma^2}\right) = \frac{N}{\sigma^2}. \tag{4.34}$$

This analysis confirms (i) the assertion from Chapter 2 that the standard error for N repeat measurements is $\alpha = \dfrac{\sigma_{N-1}}{\sqrt{N}}$, where we use σ_{N-1}, the sample standard deviation, as our best estimate of the population standard deviation; and (ii) the standard error is equivalent to the standard deviation of the mean.

Chapter summary

- When propagating the error in a measurement of A through a nonlinear function, $Z = f(A)$, the uncertainty in Z is a function of both A and its uncertainty, α_A.
- For a single-variable function the error in Z for a particular measurement $\bar{A}$ is $\alpha_Z = \left| f\left(\bar{A} + \alpha_A\right) - f\left(\bar{A}\right) \right|$.
- A calculus-based approximation for the uncertainty propagation for a single-variable function is $\alpha_Z = \left| \dfrac{dZ}{dA} \right| \alpha_A$.
- For multi-variable functions the total error in Z is obtained by adding the components from each variable in quadrature (provided the variables are independent).
- The calculus-based approximation for multi-variable functions is

$$(\alpha_Z)^2 = \left(\frac{\partial Z}{\partial A}\right)^2 (\alpha_A)^2 + \left(\frac{\partial Z}{\partial B}\right)^2 (\alpha_B)^2 + \left(\frac{\partial Z}{\partial C}\right)^2 (\alpha_C)^2 + \cdots.$$

- A hybrid approach which combines the calculus and functional approaches with results from a look-up table often proves to be the most useful.
- The best combined estimate, x_{CE}, incorporating all of the available data is the sum of the weighted means $x_{CE} = \dfrac{\sum_i w_i x_i}{\sum_i w_i}$, where the weighting is $w_i = \dfrac{1}{\alpha_i^2}$; the inverse of the square of the standard error of the weighted mean is the sum of the weightings: $\dfrac{1}{\alpha_{CE}^2} = \displaystyle\sum_i \left(\dfrac{1}{\alpha_i^2}\right)$.

Exercises

(4.1) *Propagating the error through a single-variable function*
A variable is measured to be $A = 9.274 \pm 0.005$. Calculate the mean and uncertainties in Z when it is related to A via the following relations: (i) $Z = 2A$, (ii) $Z = A/2$, (iii) $Z = \frac{A-1}{A+1}$, (iv) $Z = \frac{A^2}{A-2}$, (v) $Z = \arcsin\left(\frac{1}{A}\right)$, (vi) $Z = \sqrt{A}$, (vii) $Z = \ln\left(\frac{1}{\sqrt{A}}\right)$, (viii) $Z = \exp\left(A^2\right)$, (ix) $Z = A + \sqrt{\frac{1}{A}}$, (x) $Z = 10^A$.

(4.2) *Propagating the error through a multi-variable function*
Three variables are measured to be $A = 12.3 \pm 0.4$, $B = 5.6 \pm 0.8$ and $C = 89.0 \pm 0.2$. Calculate the mean and uncertainties in Z when it is related to A, B and C via the relations: (i) $Z = A + B$, (ii) $Z = A - B$, (iii) $Z = \frac{A-B}{A+B}$, (iv) $Z = \frac{AB}{C}$, (v) $Z = \arcsin\left(\frac{B}{A}\right)$, (vi) $Z = A \times B^2 \times C^3$, (vii) $Z = \ln(ABC)$, (viii) $Z = \exp(AB/C)$, (ix) $Z = A + \tan\left(\frac{B}{C}\right)$, (x) $Z = 10^{AB/C}$.

(4.3) *Comparing methods*
The relationship between the period, T, of the oscillation of a spring with a mass M attached to a spring with spring constant K is $T = 2\pi\sqrt{\frac{M}{K}}$. In an experiment T and M and their associated uncertainties are measured; show that the equation for the error in K is the

same using (a) the look-up tables, and (b) the calculus approximation.

(4.4) *Angular dependence of the reflection coefficient of light*
The intensity reflection coefficient, R, for the component of the field parallel to the plane of incidence is

$$R = \frac{\tan^2 (\theta_i - \theta_t)}{\tan^2 (\theta_i + \theta_t)},$$

where θ_i and θ_t are the angles of incidence and transmission, respectively. Calculate R and its associated error if $\theta_i = (45.0 \pm 0.1)°$ and $\theta_t = (34.5 \pm 0.2)°$.

(4.5) *Snell's law*
The angle of refraction θ_r for a light ray in a medium of refractive index n which is incident from vacuum at an angle θ_i is obtained from Snell's law: $n \sin \theta_r = \sin \theta_i$. Calculate θ_r and its associated error if $\theta_i = (25.0 \pm 0.1)°$ and $n = 1.54 \pm 0.01$.

(4.6) *The Lennard-Jones potential*
The Lennard-Jones potential, $V(r)$, is an effective potential that describes the interaction between two uncharged molecules or atoms, as a function of their separation r. It is written as

$$V(r) = -\frac{A}{r^6} + \frac{B}{r^{12}},$$

where A and B are positive constants. Experimentally it is easy to measure the two parameters r_0 and ϵ; here r_0 is the equilibrium separation of the pair, and ϵ the energy required to separate the pair from their equilibrium separation to infinity. Obtain expressions for A and B in terms of r_0 and ϵ. Given that for helium $\epsilon = 0.141 \times 10^{-21}$J, and $r_0 = 2.87 \times 10^{-10}$m, evaluate A and B. If ϵ can be determined to a precision of 1%, and r_0 can be determined to a precision of 0.5%, to what precision can A and B, respectively, be determined?

(4.7) *Experimental strategy*
The resistance R of a cylindrical conductor is proportional to its length, L, and inversely proportional to its cross-sectional area, πr^2. Which quantity should be determined with higher precision, L or r, to optimise the determination of R? Explain your reasoning.

(4.8) *Poiseuille's method for determining viscosity*
The volume flow rate, $\frac{dV}{dt}$, of fluid flowing smoothly through a horizontal tube of length L and radius r is given by Poiseuille's equation:

$$\frac{dV}{dt} = \frac{\pi \rho g h r^4}{8 \eta L},$$

where η and ρ are the viscosity and density, respectively, of the fluid, h is the head of pressure across the tube, and g the acceleration due to gravity. In an experiment the graph of the flow rate versus height has a slope measured to 7%, the length is known to 0.5%, and the radius to 8%. What is the fractional precision to which the viscosity is known? If more experimental time is available, should this be devoted to (i) collecting more flow-rate data, (ii) measuring the length, or (iii) the radius of the tube?

(4.9) *Error spreadsheet for van der Waals calculation*
Construct a spreadsheet which has the data from the calculation in Section 4.2.2. Include cells for: (i) the variables (molar volume and the absolute temperature), (ii) the uncertainties, and (iii) the universal gas constant as well as the parameters a and b. Verify the numbers obtained in the worked example. Repeat the calculation for (i) $V_m = (2.000 \pm 0.003) \times 10^{-3}$ m^3 mol^{-1} and $T = 400.0 \pm 0.2$ K; (ii) $V_m = (5.000 \pm 0.001) \times 10^{-4}$ m^3 mol^{-1} and $T = 500.0 \pm 0.2$ K. Repeat the calculations with the same variables for (a) He with $a = 3.457\ 209 \times 10^{-3}$ m^6 mol^{-2} Pa, and $b = 2.37 \times 10^{-5}$ m^3 mol^{-1}; (b) CO_2 with $a = 3.639\ 594 \times 10^{-1}$ m^6 mol^{-2} Pa, and $b = 4.267 \times 10^{-5}$ m^3 mol^{-1}; and (c) Ar with $a = 1.362\ 821\ 25 \times 10^{-1}$ m^6 mol^{-2} Pa, and $b = 3.219 \times 10^{-5}$ m^3 mol^{-1}.

(4.10) *Weighted mean*
A group of six students make the following measurements of the speed of light (all $\times 10^8$ m s^{-1}): 3.03 ± 0.04, 2.99 ± 0.03, 2.99 ± 0.02, 3.00 ± 0.05, 3.05 ± 0.04 and 2.97 ± 0.02. What should the cohort report as their combined result? If another student then reports $c = (3.0 \pm 0.3) \times 10^8$ m s^{-1}, is there any change to the cohort's combined measurement? If a further student reports $c = (4.01 \pm 0.01) \times 10^8$ m s^{-1}, is there any change to the cohort's combined measurement?

Data visualisation and reduction

5

The graphical representation of data is the most efficient method of reporting experimental measurements in the physical sciences. Graphs are a very effective visual aid, and we make use of them to (i) highlight trends in, and relationships among, experimental data, (ii) test theories, (iii) enable comparisons between data sets to be made, (iv) look for evidence of systematic errors and (v) extract additional parameters which characterise the data set. We treat each of these concepts in this chapter, and also emphasise the role graphs can play with regard to helping to minimise errors.[1]

5.1 Producing a good graph

The overriding considerations when producing a good graph are simplicity and clarity. To this end there are several conventions that are usually followed. A set of guidelines to follow is provided here, together with a fuller discussion of some of the points raised.

[1]Note that we assume here that a suitable computer package is being used to generate graphs, although most of the advice will also be relevant for hand-drawn graphs.

Guidelines for plotting data

(1) Plot the *independent* variable on the horizontal axis, and the *dependent* variable on the vertical axis.

(2) Consider *linearising* the data to generate a straight-line plot.

(3) Use appropriate *scales for the axes* such that most of the area of the graph is utilised.

(4) *Label each axis* with the name and units of the variable being plotted.

(5) Add *data points and error bars*, ensuring that they are clear, with different data sets being distinguishable.

(6) Add a *fit or trend line*—either a straight line, a smooth curve to capture the trend of the data set, or a suitable theoretical model.

(7) Add an informative *title* to lab-book graphs, write a *caption* for figures for publication.

A good graph enables the reader to absorb quickly the key points of the data. As such it needs to be simple, clear and contain all pertinent information. We distinguish between two instances where graphs are used extensively. Generating a **lab-book graph** by plotting the experimental data in your lab book as

Table 5.1 Prefixes used in the SI system. Note that the symbols for factors which are smaller than 1 are all lower case, and that the symbols for factors greater than or equal to a million are all upper case.

Factor	Prefix	Symbol
10^{-24}	yocto	y
10^{-21}	zepto	z
10^{-18}	atto	a
10^{-15}	femto	f
10^{-12}	pico	p
10^{-9}	nano	n
10^{-6}	micro	μ
10^{-3}	milli	m
10^{-2}	centi	c
10^{-1}	deci	d
10^{1}	deca	da
10^{2}	hecto	h
10^{3}	kilo	k
10^{6}	mega	M
10^{9}	giga	G
10^{12}	tera	T
10^{15}	peta	P
10^{18}	exa	E
10^{21}	zetta	Z
10^{24}	yotta	Y

you progress is excellent laboratory practice. The dissemination of the results of an experiment are published either in a report or a scientific paper. The graphs which appear as figures in these documents are presented in a slightly different format from the lab-book graphs and we also include guidelines to produce **graphs for publication**.

5.1.1 The independent and dependent variables

It is a convention that the data are plotted with the **independent variable** (the parameter that the experimenter is varying) on the horizontal, or x, axis and the **dependent** variable (the parameter that is measured) on the vertical, or y, axis. The x-coordinate is also called the **abscissa**, and the y-coordinate referred to as the **ordinate**.

5.1.2 Linearising the data

For clarity of the graphical representation of data one should always attempt to show a **linear relationship** between the dependent and independent variables. This is because it is much easier to (i) see deviations from a straight line, (ii) fit linear relations and express the relationship between the experimental quantities and those predicted by theory. In particular there exist analytic expressions for the slope and intercept and their uncertainties for a straight-line fit. For example, it is known that the period, T, of a simple pendulum depends on the length L via the relation $T = 2\pi\sqrt{\frac{L}{g}}$, where g is the acceleration due to gravity. Typically one sets the length of the pendulum (the independent variable) and makes multiple measurements of the period (the dependent variable) to give $T \pm \alpha_T$. Thus by plotting T^2 on the y-axis versus L on the x-axis we should get a straight line through the origin, and can extract a value for g and its error from the slope. Note that the conventional terminology is to call this is 'a graph of T^2 against L'.

5.1.3 Appropriate scales for the axes

It is important that the range of each axis is independently adjusted such that the data set is fully encapsulated without large areas of the graph being empty. The default settings in most graphical packages try to achieve this goal; however to produce a clear graph the scales may need to be optimised further. A common problem is in defining the minimum value of the axis range—think carefully whether an axis should start at 0. It is possible to include a smaller figure as an inset to a larger graph if there is plenty of white space available—this is an efficient way to convey more information in the same space (a vital consideration when preparing figures for publication in prestigious journals, or laboratory reports for assessment when there is a strict page limit).

5.1.4 Labelling the axes

Each axis must be labelled with the variable being plotted (either the name, or the accepted symbol) and units. There are two extensively used conventions as to how units are included: one can either (i) write the unit in parentheses after the variable, e.g. displacement (m); or (ii) the unit can be separated from the quantity with a solidus (/), e.g. displacement/m. Use SI units wherever possible. When plotting data the use of arbitrary units should be avoided as much as possible; if this is impossible the correct terminology to label the axis is (arbitrary unit) not (a.u.) as this can be confused with, for example, astronomical unit or atomic unit. To improve the clarity of a graph, use of exponential notation or numbers containing many decimal places should be avoided. This can be achieved by using appropriate multipliers and prefixes. The standard factors range from 10^{-24} to 10^{24} and are listed in Table 5.1; for example 0.000 005 A and 5E − 6 A both become 5 μA.

Many different data sets can be plotted simultaneously on one graph in order to establish trends among different parameters, see Fig. 5.1 for an example of current–voltage characteristics for three different semiconductor diodes. When plotting multiple data sets on a single graph different and distinguishable symbols should be used for distinct sets. An explanation of which symbol represents which data set can either be added as an inset to the graph, or explained clearly in the caption as is demonstrated in the caption of Fig. 5.1 (some journals will specify a preference). The information from the caption of Fig. 5.1 could also be conveyed by the sentence 'the open squares are for germanium, the open circles for silicon, and the solid stars for the zener diode'.

5.1.5 Adding data points and error bars to graphs

Experimental data are plotted as discrete points using a common symbol for a single data set. Most graphical plotting packages have built-in symbols that are distinct, scalable and easily distinguishable. The symbol size should be adjusted for maximum clarity. When the number of data points is large, such as occurs with computer-recorded spectra, it may be more appropriate to plot the data as continuous lines without a symbol as even the smallest symbols would overlap and obscure the underlying trend. If plotting several of these data sets on a single graph it may be useful to plot a subset of the points on the curves to differentiate between the different data sets and enhance the clarity of the figure. As in Fig. 5.1 a common and unique symbol should be used for each different data set. For lab-book graphs it is important that a legend detailing the different data sets plotted is included within the area of the graph; publication graphs would have this information either within the area of the graph, or in a detailed caption under the figure.

The coordinates of a data point on a graph are our best estimate (i.e. the mean) of the independent and dependent variables. There is an uncertainty associated with these mean values and a graph needs to reflect this. In Chapter 2 we showed how the standard error of the mean, α, is calculated for single variables, and in Chapter 4 how these could be propagated through various functions. We represent the final error in both the abscissa and ordinate as an error bar which is the two-thirds confidence limit that the measured value, $\overline{Z}$, lies within the range $(\overline{Z} - \alpha_Z^-)$ to $(\overline{Z} + \alpha_Z^+)$, as shown in Fig. 5.2. To produce a straight-line graph, we often process the data. In this case, although the uncertainty in the original measurements might be equal and symmetric,

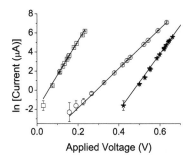

Fig. 5.1 Combining many different data sets on one graph. The forward-bias current–voltage characteristics of a silicon (o), germanium (□) and zener (⋆) diode are depicted. A logarithmic scale was chosen for the current to emphasise the exponential growth of the current with increasing voltage.

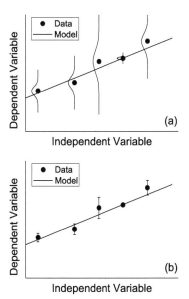

Fig. 5.2 Error bars are added to points on a graph to indicate the 68% confidence limits. The Gaussian distributions shown in (a) represent the probability distribution function for the mean value, and we expect 68% of the points to be within one standard error of the mean of this value. In (b) we represent the same information as in (a), but plot error bars with a magnitude of one standard deviation of the distributions shown in (a).

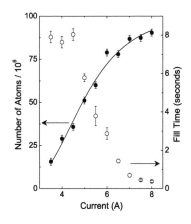

Fig. 5.3 The number of atoms (in millions) and the lifetime of the atoms in a magneto-optical trap are displayed as a function of the dispenser current. By plotting both dependent variables simultaneously the trade-off between a high number of atoms and long lifetime is evident. The smooth curve is a guide to the eye.

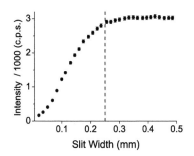

Fig. 5.4 The intensity of X-rays transmitted as a function of the slit width.

by propagating through a nonlinear function the uncertainties vary from point to point and may also be asymmetric. In general, the fractional errors on the dependent variable should be larger than those of the independent variable— the analysis in subsequent chapters assumes that this is the case. A discussion of how to proceed when both variables contain errors is given in Chapter 9. If the error bars are not larger on the vertical axis than those on the horizontal axis consider revising the experimental strategy.

Horizontal and vertical error bars should be plotted as long as they do not obscure the clarity of the graph. A common issue is when the size of the symbol used to plot the data is larger than the size of the error bar. In this case, consider reducing the size of the symbol within the constraint of graph clarity. If the measurements are precise the error bars may be too small to be seen, as occurs in Fig. 5.10 below. This should be noted in the figure caption on published graphs and included as an annotation in lab-book graphs. If the trend in the data is clear as there are many data points, it may reduce the clarity of the graph to add every error bar. Representative error bars can be added to a small subset of the data in this case.

Plotting error bars

The uncertainty in a quantity is represented graphically using error bars. The error bars are drawn between $\left(\overline{Z} - \alpha_Z^-\right)$ and $\left(\overline{Z} + \alpha_Z^+\right)$, where α_Z is the standard error. For Poisson statistics with N counts the error bar is drawn between $\left(N - \sqrt{N}\right)$ and $\left(N + \sqrt{N}\right)$.

There are circumstances in which two dependent variables are measured for each value of the independent variable. In such a situation it is possible to draw a graph with two y-axes, provided a sufficient level of clarity is maintained. Figure 5.3 shows an example of data obtained for the number and lifetime of cold atoms in a trap as a function of the current passed through the atom dispenser. It is crucial to indicate clearly which data set is associated with which axis. A smooth curve can be added to emphasise the trend in the data where there is no particular theoretical model, with an explanation in the caption that the curve is a guide to the eye.

5.1.6   Adding a fit or trend line

There are three types of fit, or trend line, which we can add to a graph. (1) If a visual inspection of the graph confirms a linear dependence, we add a **linear trend line**. This topic is discussed in detail in Section 5.2.1. (2) There are occasions when the theoretical model cannot be linearised. For such a case the appropriate nonlinear function can be added to the graph. The relevant parameters are initially chosen, by eye, to capture the trend of the data. There will be an extensive discussion in Chapters 6 and 7 of how to optimise the parameters and evaluate their uncertainties. (3) In circumstances where there is no theoretical prediction of the relationship between the dependent and independent variable one can add a smooth curve to guide the eye. The caption should clearly state that the smooth curve is a guide to the eye, and not a theoretical prediction.

5.1.7 Adding a title or caption

For a lab-book graph an informative title should be included. Graphs for a publication would have this information in a detailed caption under the figure. Graphs in this book have captions, not titles. To improve clarity, the use of shading, coloured backgrounds and grid lines should be avoided.

5.2 Using a graph to see trends in the data

It is far easier to see trends in a data set from a graph rather than the raw data—this is why it is important to plot graphs as data are being collected. Consider the data shown in Fig. 5.4 which shows the intensity of X-rays transmitted as a function of the slit width. The graph allows us to deduce that (a) for widths smaller than approximately 0.25 mm an increase in slit width has a concomitant increase in transmitted X-ray intensity: (b) for widths greater than approximately 0.25 mm the X-ray intensity is largely independent of width. Both of these statements can be put on a more quantitative footing later, but the graphical representation of the data allows us to draw preliminary conclusions rapidly.

In Fig. 5.5(a) the number of laser-cooled Cs atoms in a magneto-optic trap is plotted as a function of time, showing a monotonic decrease. Theoretical considerations lead us to believe that the dependence might be an exponential decay, which motivated plotting the natural logarithm of the atom number, as shown in parts (b) (using a logarithmic scale) and in part (c) by evaluating the logarithm of the dependent variable. For the logarithmic plots we see that there are two different straight lines: the first is associated with a time constant of $\tau = 8.67 \pm 0.04$ s, and the second with a time constant of $\tau = 67 \pm 4$ s. Further theoretical analysis leads us to believe that the former is associated with light-assisted collisions, and the latter with collisions with background gas molecules. Most of this information was gleaned rapidly through plotting the appropriate graph. It is possible to extract values for the two decay constants and their uncertainties from the straight-line fits to such a trace; it is also possible by repeating the experiment many times to obtain many readings of the time constants, and the results of Section 4.5 can be used to combine these independent measurements.

5.2.1 Adding a linear trend line

Experiments are often designed such that the dependent variable exhibits a linear dependence on the independent variable. There are simple analytic results for calculating the best-fit straight line for a data set with constant error bars on the vertical axis (we will discuss in more detail in the next section, and extensively in later chapters, what exactly we mean by 'best fit'). Most

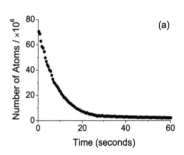

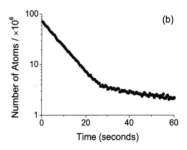

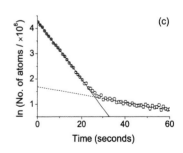

Fig. 5.5 The number of atoms in a trap as a function of time. In (a) the number is plotted against time, in (b) a logarithmic scale is chosen for the *y*-axis, and in (c) the natural logarithm of the atom numbers is plotted against time for every third point shown in (a) and (b).

graphical packages and spreadsheets allow you to add a linear trend line, which is the calculated best-fit straight line to the data.[2]

[2] We assume that a computer plotting package is being used to add the trend line, therefore we will not give advice about using a clear plastic ruler, using a sharp pencil, etc. which is relevant for a hand-drawn graph.

As was discussed in Section 5.1.5 the error bars on the y-axis of a graph represent the 68% confidence limit for the value of the ordinate. If we assume that the best-fit straight line is a good description of the data, we would therefore expect that the line intersects two-thirds of the measurements within their error bar.

For a **good fit** we expect two-thirds of the data points to be within one standard error bar of the trend line.

There are two obvious reasons why the fraction of points which are consistent with the line of best fit may be significantly different from this: (i) the error bars have been overestimated; and (ii) the assumed linearity of the best fit is not valid. A thorough discussion of whether a model (be it a linear fit or otherwise) describes a data set is reserved until Chapter 8.

If approximately two-thirds of the data points intersect the linear trend line, the data are well modelled by a straight line $y = mx + c$, and it is appropriate to extract four important parameters—these are the gradient, m, and intercept, c, and their associated errors.[3] Using a process known as the **method of least squares** (to be discussed in the next section) one can derive analytic expressions for the gradient and its uncertainty and the intercept and its uncertainty (Taylor 1997, pp. 182–8, or Barford 1985, Section 3.3). The results are:

[3] In Chapter 7 we will argue that the correlation coefficient should also be reported for a straight-line fit.

$$c = \frac{\sum_i x_i^2 \sum_i y_i - \sum_i x_i \sum_i x_i y_i}{\Delta}, \tag{5.1}$$

and

$$m = \frac{N \sum_i x_i y_i - \sum_i x_i \sum_i y_i}{\Delta}, \tag{5.2}$$

with the error in the intercept,

$$\alpha_c = \alpha_{CU} \sqrt{\frac{\sum_i x_i^2}{\Delta}}, \tag{5.3}$$

and the error in the gradient,

$$\alpha_m = \alpha_{CU} \sqrt{\frac{N}{\Delta}}, \tag{5.4}$$

where[4]

[4] With a carefully constructed spreadsheet the terms $\sum_i x_i$, $\sum_i x_i^2$, $\sum_i y_i$, $\sum_i x_i y_i$, and $\sum_i (y_i - mx_i - c)^2$ for the evaluation of eqns (5.1)–(5.6) are easily determined. Note also that many spreadsheets and software packages have in-built functions which perform these summations.

$$\Delta = N \sum_i x_i^2 - \left(\sum_i x_i\right)^2, \tag{5.5}$$

and the so-called **common uncertainty** α_{CU} is defined as

$$\alpha_{CU} = \sqrt{\frac{1}{N-2} \sum_i (y_i - mx_i - c)^2}. \tag{5.6}$$

The common uncertainty will be discussed in greater detail in Chapter 8—essentially it represents an approximation to the magnitude of the uncertainty in the y measurements, assumed to be common to all data points, assuming that the straight line is the appropriate theoretical model to describe the data.

Note that the gradient and intercept are parameters and should be quoted using the five golden rules introduced at the end of Chapter 2. Note also that quoting the slope and intercept is an example of **data reduction**. In the same way that we introduced three numbers (the mean, standard deviation and standard error) to summarise the information contained within N data points, the four numbers (slope, intercept and their uncertainties) are an attempt to summarise the information from N data pairs.

5.2.2 Interpolating, extrapolating and aliasing

By adding a trend line to a discrete data set we have some confidence in an underlying model. However, we must be very careful in over-interpreting the data, as, at a fundamental level, all we have are our pairs of discrete measurements of the dependent and independent variables, and their uncertainties.

The process of using the straight line (or any other smooth-curve fit to a data set) to infer a value of the dependent variable for a value of the independent variable between two measured values is called **interpolating** and is shown in Fig. 5.6. The process of predicting a value for the dependent variable when the independent variable is outside the measured range is called **extrapolating**; both processes should be used with caution. There are many examples of physical phenomena where the linear dependence exhibited for small values of the independent variable breaks down at higher values: for example, a spring shows a linear extension as a function of the applied force only up to the elastic limit; the number of atoms or molecules in an excited state increases linearly with laser intensity initially, but saturates at higher intensities.

If the values chosen for the independent variable are periodic it is possible to suffer from the problem of **aliasing**. For example, if voltages in an a.c. electrical circuit are sampled at times $t = 0, 1, 2, \ldots$ seconds, the three waveforms $V_1(t) = 5$ volts, $V_2(t) = 5\cos(2\pi \times t)$ volts, and $V_3(t) = 5\cos(2\pi \times 3t)$ volts will all give identical results; the time-dependent functions are said to alias the constant. If measurements of temperature which are evenly spaced in time give the results $21.1°C$, $21.1°C$, $21.2°C$, $21.3°C$ and $21.4°C$, what temperature might we expect half-way between the last two measurements? We could fit a straight line, or a higher order polynomial, to aid with the interpolation. However, if we learn that the measurements were of the temperature in a particular city at midday, the measurements will be useless for predicting the temperature at midnight.

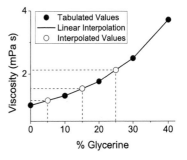

Fig. 5.6 Five tabulated values of the viscosity of an aqueous glycerol solution as a function of the percentage of glycerine are plotted (solid dots). It is reasonable to expect the viscosity to vary smoothly as a function of the percentage of glycerine; thus one can interpolate to estimate the viscosity at values other than the tabulated ones. The open circles show the results of a linear interpolation between successive points for 5, 15 and 25% glycerine. A higher-order polynomial fit could also be used if required.

5.3 Introduction to the method of least squares and maximum likelihood

Most straight-line graphs we draw have a large number of data pairs, and we wish to learn about the slope and intercept. The situation under consideration

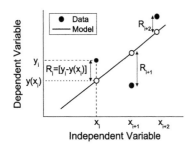

Fig. 5.7 The residual of a data point is defined as the measured value, y_i, minus the theoretical y-value for that x-value. In the method of least squares the sum of the squares of the residuals is minimised by changing the value of the slope and intercept of the fitted line.

is one that mathematicians call **overdetermined**, i.e. we have more equations than unknowns. To extract two parameters (slope and gradient) is obviously impossible with one data point; we could solve for both m and c (with no uncertainty) if we had two *exact* pairs of data (x_1, y_1) and (x_2, y_2). The situation we face is very different: we typically have far more than two data points, but the experimental values are subject to random statistical fluctuations. We therefore perform what is known as **regression analysis**, where the goal is to determine the optimal values of parameters for a function (a straight line in the present case) that lead to the function having the best fit to the set of experimental data.

The **best-fit straight line** is defined as the line which is in closest proximity to as many of the data points as possible. If the coordinates of the ith data point are (x_i, y_i) and the line of best fit is of the form $y = mx + c$, then the y-coordinate of the best-fit line at x_i is $y = mx_i + c$. If we have selected appropriate values of m and c, then the difference $(y_i - y)$ will be small—this difference between the experimentally measured value of the dependent variable and the theoretical prediction is called the **residual**. The definition of the residual is shown schematically in Fig. 5.7. We cannot simply obtain the best-fit line by minimising the difference, $(y_i - y)$, for all data points as sometimes the residual will be negative, and sometimes positive (Fig. 5.7). Instead, we must consider a quantity which is always positive. One approach is to minimise the modulus of the difference, $|y_i - y|$; an alternative, and widespread, approach is to minimise the square of the difference, $(y_i - y)^2$, which is also always positive. In the **method of least squares** one seeks to minimise the sum of the squares of the residuals. By definition, therefore, the best values of m and c will be those in which the squares of the differences summed for all data points is minimised. The method of least squares can be derived from the formalism of **maximum likelihood** in conjunction with the central limit theorem (see Section 3.5), which motivates the statement that each data point that we measure, y_i, is drawn from a Gaussian distribution with a width given by the standard error, α_i. The link between the Gaussian distribution and the error bar was highlighted in Fig. 5.2.

The y-coordinate of the line of best fit at x_i, $y(x_i)$, is the most probable value of the mean of the parent distribution. We can use the parent distribution to calculate the probability of obtaining the value y_i, given the parameters m and c, which is proportional to the value of the probability density function at y_i. The assumption that we make is that the parent distribution is described by the Gaussian probability density function given in Section 3.2, eqn (3.7):

$$P_i = G_i \, \mathrm{d}y_i = \frac{1}{\sqrt{2\pi}\,\alpha_i} \exp\left[-\frac{(y_i - mx_i - c)^2}{2\alpha_i^2}\right] \mathrm{d}y_i. \tag{5.7}$$

The total probability that we obtain our observed set of N measurements given m and c is the product of the probabilities for each individual measurement:

$$P(m, c) = \prod_i P_i = \prod_i \frac{\mathrm{d}y_i}{\sqrt{2\pi}\,\alpha_i} \exp\left[-\frac{1}{2}\sum_i \left(\frac{y_i - mx_i - c}{\alpha_i}\right)^2\right]. \tag{5.8}$$

In eqn (5.8) we think of the values of y_i, α_i and x_i as being fixed once the experiment is complete, and of m and c as variables whose values can be chosen to maximise the probability, $P(m, c)$—which occurs when the line of best fit is as close as possible to the data points. The analytic solutions to the line of best fit presented in eqns (5.1)–(5.6) were found by differentiating $P(m, c)$ with respect to m and c and setting the result to zero, for the special case of all the uncertainties α_i being equal. In eqn (5.8) the pre-factor is independent of m and c and the probability is maximised when the summation term in the argument of the exponent is minimised. The probability is maximised when the **goodness-of-fit** parameter, χ^2, is minimised. We have defined χ^2 as:

$$\chi^2 = \sum_i \left(\frac{y_i - y(x_i)}{\alpha_i} \right)^2 . \tag{5.9}$$

When the error bars, α_i, are constant they do not influence the values of the best-fit intercept, gradient or their associated errors. We shall see in Chapter 6 that, in general, to find the line of best fit we vary the parameters m and c to minimise the weighted sum of the square of the deviations, χ^2.

5.3.1   Example using the method of least squares

As a simple example, consider the case shown in Fig. 5.8. The data points obviously show a linear trend through the origin. In this, rather contrived, example we know that $y = mx$. But how do we arrive at the best value for m? One method is to try many different values of m until we see that the hypothesised model goes through our data points (Fig. 5.8a). By such a method it becomes obvious that the best line is $y = 2x$.

Alternatively, we can arrive at the same result by minimising the sum of the squares of the residuals, $\sum_i (y_i - y)^2$. The best-fit line will be the one with the value of m for which the sum of the squares of the residuals is a minimum. By inspection of the plot of $\sum_i (y_i - y)^2$ against m (Fig. 5.8b) we see the well-defined minimum (in this case zero) corresponding to the best fit with $m = 2$. Obviously, with genuine experimental data, the minimum value of $\sum_i (y_i - y)^2$ will never be zero for two reasons: (i) the inevitable statistical fluctuations associated with the data, and (ii) the hypothesised model (a straight line in this case) might not be an appropriate description of the data over the entire range.

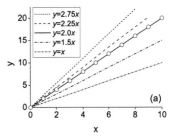

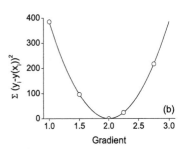

Fig. 5.8 The evolution of the goodness-of-fit parameter with the gradient, m. The open points in (b) correspond to the lines shown in (a).

5.4   Performing a least-squares fit to a straight line

There are three methods for obtaining the best-fit straight line parameters for a given a data set. Firstly, the analytic functions of eqns (5.1)–(5.6) can be used to determine both the value and uncertainty of the slope and intercept. The terms to be evaluated within these equations are commonly found on scientific calculators and spreadsheet analysis packages. Secondly,

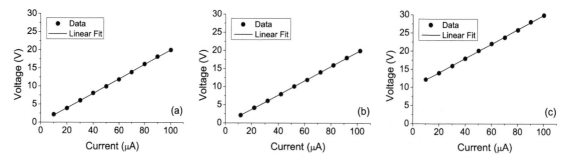

Fig. 5.10 Data from three experiments used to determine the value of a resistor. Error bars too small to be seen. In (a) a correctly calibrated voltmeter and ammeter were used giving a fitted gradient of 200 ± 2 kΩ and an intercept, consistent with Ohm's law, of 0.0 ± 0.1 V. In (b) the data are recorded with an ammeter with a zero error. The gradient remains the same as (a), but the intercept is -0.26 ± 0.06 V. In (c) the voltmeter has a calibration error—again the gradient remains the same as in (a) with an intercept of 10.09 ± 0.09 mV. In both (b) and (c) the intercept is not consistent with Ohm's law and indicates a systematic error. Note the axes scales have been set to be the same for these three graphs to highlight the differences between the curves.

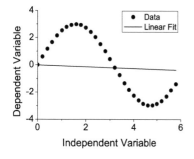

Fig. 5.9 The least-squares best-fit straight line to a sinusoidal variation. Error bars smaller than symbol size. **Health warning** It is possible to use the method of least squares to find the 'best-fit' straight line for *any* data set. This *does not* mean that a linear fit is an appropriate theoretical model.

any computer fitting/plotting software will have the ability to conduct a least-squares regression. Finally, one could construct the appropriate terms within a spreadsheet and use the in-built minimisation routine. The last method may be over-elaborate for this situation; however it is easily generalised to far more complex fitting and analysis, and is used extensively later in the book.

The method of least squares will *always* fit a straight line to a data set, but does not answer another interesting question, namely 'are the data consistent with a straight line?'. One could 'fit' a straight line to a data set which exhibits sinusoidal variation, as seen in Fig. 5.9, with obvious ridiculous consequences if one takes the 'best-fit' parameters seriously. We postpone until Chapter 8 the mathematically rigorous techniques for giving a quantitative measure for the quality of a fit, but note that, in Fig. 5.9, 68% of the data points do not coincide with the line of best fit and there would be little point in continuing with an analysis based on the least-squares best-fit straight line.

5.5   Using graphs to estimate random and systematic errors

A single pair of measurements of current and voltage through a resistor allows us to calculate an experimental value of the resistance and its error. This assumes that Ohm's law is valid for the particular current and voltage parameters used. For example, for a voltage of 6.0 ± 0.1 V and a current of 30.0 ± 0.1 μA, we deduce a resistance, R, of 200 ± 3 kΩ. A better experimental strategy is to take a series of measurements for different current–voltage pairs, and plot a graph, as shown in Fig. 5.10(a). Any departure from the linear trend is visually immediately apparent. The best estimate of the resistance is the gradient of the graph, which we can extract by fitting a least-squares straight line. By collecting more data the statistical error in the resistance will be reduced.

Imagine the meter we use to measure the current has a systematic error, and each of the current values is systematically too high by 2μA. The single-point measurement of 6.0 ± 0.1 V and a current of 32.0 ± 0.1 μA, yields a resistance of 188 ± 3 kΩ, which is incorrect. However, by plotting the graph the fact that *all* of the data points are systematically displaced will have no bearing on the slope of the graph, and hence our determination of the resistance; see Fig. 5.10(b). Similarly, a systematic offset on all of the voltage readings will give an incorrect single-point measurement, but the slope of the graph will be correct, as in Fig. 5.10(c). Thus a whole class of systematic errors can be negated by plotting a graph. In both of these cases, the intercepts are not consistent with zero—a sign of the presence of a systematic error.

5.6 Residuals

The best-fit line in Fig. 5.9 is clearly not a good description of the data. Here we introduce the idea of performing a preliminary test of the quality of the fit from the best-fit straight line. Based on the ideas discussed in Section 5.3 and Chapter 3 we would only expect two-thirds of the data points to lie on the best-fit line, within their error bars. Therefore a quick visual inspection of a line of best fit to a graph should be performed; approximately two-thirds of the data points should be consistent with the line, and approximately half of the other points should lie above the line, approximately half below.

It is very useful to plot the residuals $R_i = y_i - y(x_i)$. If the data are consistent with a straight line, the residuals should have a mean of zero and show no obvious structure. If there are many data points, a histogram of the residuals can be plotted. One would expect to see a Gaussian distribution centred on zero. The power of plotting the residuals is evident in the two graphs of Fig. 5.11. A quick visual inspection in both cases indicates that a linear model

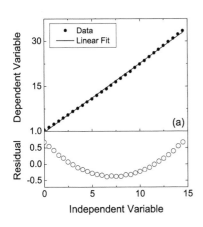
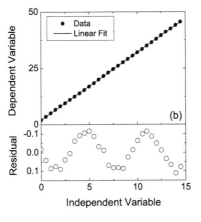

Fig. 5.11 Data sets with a linear fit. The error bars are too small to be seen. In both cases, a quick visual inspection suggests a good fit. However, a plot of the residuals shows structure. In (a) we clearly see that a quadratic terms needs to be incorporated into the model of our data and in (b) a sinusoidal variation should be included.

seems a good fit; however, in both cases, the residuals clearly indicate that higher order terms have to be included in the model.

Chapter summary

- The overriding considerations when producing a good graph are simplicity and clarity.
- The error bar on a point on a graph is the two-thirds confidence limit.
- The error bar is drawn between $\left(\overline{Z} - \alpha_{\overline{Z}}^-\right)$ and $\left(\overline{Z} + \alpha_{\overline{Z}}^+\right)$, where α is the standard error (standard deviation of the mean).
- For Poisson statistics with N counts the error bar is drawn between $\left(N - \sqrt{N}\right)$ and $\left(N + \sqrt{N}\right)$.
- The method of least squares can be used to deduce the best-fit parameters for linear regression $y = mx + c$ where the data have constant error bars.
- The best estimate of the intercept and slope are

$$c = \frac{\sum_i x_i^2 \sum_i y_i - \sum_i x_i \sum_i x_i y_i}{\Delta}, \text{ and}$$

$$m = \frac{N \sum_i x_i y_i - \sum_i x_i \sum_i y_i}{\Delta},$$

  where $\Delta = N \sum_i x_i^2 - \left(\sum_i x_i\right)^2$.
- The best estimate of the error in the intercept and slope are

$$\alpha_c = \alpha_{\text{CU}} \sqrt{\frac{\sum_i x_i^2}{\Delta}}, \text{ and } \alpha_m = \alpha_{\text{CU}} \sqrt{\frac{N}{\Delta}},$$

  where $\alpha_{\text{CU}} = \sqrt{\dfrac{1}{N-2} \sum_i (y_i - mx_i - c)^2}$.
- Having plotted a straight-line graph the residuals should be analysed—two-thirds of the data points should be consistent with the line within their error bars.

Exercises

(5.1) *Geometry of the best-fit straight line*
Show, using eqns (5.1) and (5.2), that the line of best fit goes through the point $(\overline{x}, \overline{y})$.

(5.2) *Linearising functions for plotting*
How can the following functions be linearised, suitable for a straight-line plot? Explain what transformation is needed for the functions to get them to be of the form $y = mx + c$. What would the slope and intercept of the line be? Take U to be the independent variable, and V the dependent variable.

(i) $V = aU^2$,

(ii) $V = a\sqrt{U}$,

(iii) $V = a \exp(-bU)$,

(iv) $\frac{1}{U} + \frac{1}{V} = \frac{1}{a}$.

(5.3) *Best-fit straight line—an unweighted fit*

The data listed below come from an experiment to verify Ohm's law. The voltage across a resistor (the dependent variable) was measured as a function of the current flowing (the independent variable). The precision of the voltmeter was 0.01 mV, and the uncertainty in the current was negligible.

Current (μA)	10	20	30	40	50
Voltage (mV)	0.98	1.98	2.98	3.97	4.95
Current (μA)	60	70	80	90	
Voltage (mV)	5.95	6.93	7.93	8.91	

(i) Use the results of Section 5.2.1 to calculate the unweighted best-fit gradient and intercept, and their uncertainties. (ii) Calculate the common uncertainty, α_{CU}, and compare the value with the experimental uncertainty. (iii) Plot a graph of the data and add the best-fit straight line. (iv) Calculate the residuals, and comment on their magnitudes.

Least-squares fitting of complex functions

Often the data we wish to plot either have error bars which are not uniform, or the data are described by a function which is not a simple straight line. In both cases, the graphical representation of the data remains the most efficient method of highlighting trends, testing our theories and enabling comparisons between data sets to be made. However, if one wishes to fit the data, we need to extend our strategies beyond those discussed in the previous chapter.

6.1 The importance of χ^2 in least-squares fitting

The dimensionless quantity χ^2 defined in Chapter 5 as a goodness-of-fit parameter,

$$\chi^2 = \sum_i \frac{(y_i - y(x_i))^2}{\alpha_i^2}, \tag{6.1}$$

is also valid for non-uniform error bars. In Chapter 5 we treated the special case of the error bars, α_i, all being equal and the function $y(x_i)$ being the straight line $y(x_i) = m x_i + c$. In this chapter we will relax these constraints, and consider the more general case. First, we will discuss the situation where the errors are no longer uniform and in the later part of the chapter we will consider in detail examples of fits to more complex functions. In all cases, the best-fit parameters remain those for which χ^2 is minimised. Note that, being a function of the random variables y_i, χ^2 is in turn a random variable with a probability distribution function and we return to this concept in Chapter 8.

There are three methods for obtaining the best-fit parameters to a given data set. Firstly, for some simple theoretical models (such as a straight line or low-order polynomials) analytic functions can be used to determine the value of the best-fit parameters such as slope and intercept. However, the two disadvantages of this method are that (i) there are no easily accessible closed-form expressions to calculate the uncertainties in the parameters, and (ii) it is impossible to generalise the analytic results to an arbitrary function. Secondly, most computer-based fitting/plotting software will have the ability to conduct a weighted least-squares regression which will give the best-fit parameters and their associated uncertainties. Finally, one could perform the appropriate analysis in a spreadsheet and use the in-built routine to minimise

χ^2. For all three cases, the final minimised value of χ^2, χ^2_{min}, can then be used to determine the quality of the fit.

There are three related questions when considering the 'best fit': (i) are my data consistent with the proposed model at a particular confidence level? Assuming the model is valid, (ii) what are the best-fit parameters of the model? and finally (iii) what are the uncertainties in these best-fit parameters? The first question is often the hardest to answer, and we return to it in Chapter 8. Throughout this chapter we implicitly assume that the uncertainties in our data set are Gaussian and that the proposed theoretical model is a valid description of the data. We use the phrase 'good fit' for the case where there is agreement between our data and the proposed theoretical model.

6.1.1 χ^2 for data with Poisson errors

When the sample distribution is a discrete function, i.e. where the experiment involves counting, we have seen that the distribution of the measurements is given by the Poisson probability distribution function with a mean count O_i and associated uncertainty $\alpha_i = \sqrt{O_i}$ (Section 3.4). Here O_i is the **observed** number of counts for the ith interval.

From the definition of χ^2 one may think that for Poisson statistics we should substitute $\alpha_i = \sqrt{O_i}$ into eqn (6.1). However the appropriate formula for χ^2 for Poisson statistics (see Squires 2001, Appendix E, or Taylor 1997, Chapter 12) is:

$$\chi^2 = \sum_i \frac{(O_i - E_i)^2}{E_i},\tag{6.2}$$

where E_i is the **expected** number of counts in the same interval. To ensure that the χ^2 calculation is not skewed by any asymmetry in the Poisson probability distribution function for low means, it is important that the data are re-binned such that within each re-binned interval the sum of the expected counts is not too low; the threshold is usually set as five. Note that if we have a good fit, then by definition, $\alpha_i = \sqrt{E_i} \approx \sqrt{O_i}$.

Equation (6.2) **should only be used for Poisson counts**, as it is a special case of eqn (6.1). Applying eqn (6.2) to any other situation leads to nonsensical results because χ^2 will no longer be dimensionless.

6.2   Non-uniform error bars

A sequence of data points with non-uniform error bars is referred to as **heteroscedastic**. There are many reasons why the data we obtain and wish to plot may have error bars which are not uniform; these may include:

- The uncertainty in a variable may be intrinsically non-uniform; this occurs with, for example, a Poisson distribution of counts in radioactive decay, where the errors decrease as the count rate decreases (although the fractional error increases).

- Different numbers of repeat measurements have been taken for each value of the independent variable.
- Measuring devices may have similar percentage precision for different scale settings. The magnitude of the error bar will then depend on the scale selected. (Multimeters are a common example of this—the precision of the instrument could be 0.1 μA on the 200 μA scale but 0.1 mA on the 200 mA scale.)
- The process of linearising the data through a nonlinear function will result in non-uniform error bars for the processed data even when the uncertainty in the raw data is uniform.

Figure 6.1 shows four examples of heteroscedastic data sets. In part (a) we see a data set where the percentage error of the ordinate grows approximately linearly with the value of the abscissa. Part (b) displays the natural logarithm of the count rate as a function of time for a background-corrected radioactive decay. There are two factors which lead to the non-uniform error bars here: (i) the inherent nonlinearity of the error in Poisson counts, and (ii) the non-linearity introduced by propagating the error through the logarithm. A data set to measure the speed of light by measuring a phase shift as a function of separation between source and detector is shown in Fig. 6.1(c). As we discussed in Chapter 4, there is an inherent nonlinearity in the error in the phase deduced from the Lissajous method; note that the error is relatively large when the phase is close to $\pi/2$. The output of a frequency-to-voltage converter is seen in Fig. 6.1(d), where it is evident that when the frequency is a multiple of the mains frequency (50 Hz) the signal becomes more noisy.

6.3   A least-squares fit to a straight line with non-uniform error bars

In Chapter 5, Section 5.2.1, we applied the method of least squares to evaluate the slope and intercept (and their uncertainties) of a straight-line fit to a data set without taking into account the uncertainties in the measurements. It is possible to include the uncertainties in the analysis, and in this section we extend the treatment to perform a **weighted least-squares fit** to the straight line $y(x_i) = m x_i + c$. The method of least squares can be used to generate analytic expressions (Taylor 1997, pp. 201–4, or Press *et al.* 1992, Section 15.2) for the slope, m, the intercept, c, and their uncertainties α_m and α_c:

$$c = \frac{\sum_i w_i x_i^2 \sum_i w_i y_i - \sum_i w_i x_i \sum_i w_i x_i y_i}{\Delta'}, \tag{6.3}$$

and

$$m = \frac{\sum_i w_i \sum_i w_i x_i y_i - \sum_i w_i x_i \sum_i w_i y_i}{\Delta'}, \tag{6.4}$$

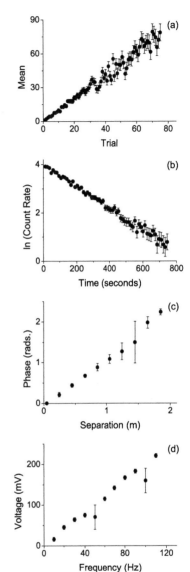

Fig. 6.1 Four examples of heteroscedastic data sets well-described by a straight line. In (a) there is a constant percentage error, in (b) Poisson counts of a radioactive decay have non-uniform error bars. In (c) the large error bar in phase near $\pi/2$ is highlighted when using the Lissajous method is highlighted, whereas (d) illustrates the degradation of the signal-to-noise ratio from a frequency-to-voltage converter near harmonics of the mains frequency.

[1]With a carefully constructed spreadsheet the terms w_i, $w_i x_i$, $w_i x_i^2$, $w_i y_i$, $w_i x_i y_i$, their products and summations required for the evaluation of eqns (6.3)–(6.7) are easily determined.

with errors,[1]

$$\alpha_c = \sqrt{\frac{\sum_i w_i x_i^2}{\Delta'}},$$

(6.5)

and

$$\alpha_m = \sqrt{\frac{\sum_i w_i}{\Delta'}},$$

(6.6)

where

$$\Delta' = \sum_i w_i \sum_i w_i x_i^2 - \left(\sum_i w_i x_i\right)^2.$$

(6.7)

[2]Equations (5.1)–(5.5) in Chapter 5 are a special case of eqns (6.3)–(6.7) with the weighting, w, being the same for each data point.

The weighting for each point, from eqn (4.25), is the inverse square of the uncertainty, $w_i = \alpha_i^{-2}$, and the summation is over all the data points.[2]

We will now illustrate the mathematical results of eqns (6.3)–(6.6) with worked examples from Fig. 6.1 and we will compare the results of the weighted fit with the unweighted values obtained in Chapter 5.

There are two fitted lines shown on both graphs in Fig. 6.2. The dashed line is the unweighted least-squares fit. This procedure gives equal weighting to all the data points. When the error bars are not uniform, a better strategy is to give more weighting to the data points with least uncertainty; this is inherent in eqns (6.3)–(6.6). The results of this weighted least-squares fitting are shown as the solid lines in Fig. 6.2. The results of the fits are summarised in Table 6.1.

For graph (a) in Fig. 6.2 the weighted and unweighted fits give very similar gradients, and errors in gradients. Note, however, the order-of-magnitude reduction in the uncertainty in the intercept obtained with the weighted least-squares fit. The weighted fit gives significantly more credence to the data with the smallest error bars; in this case these are the ones close to the origin, hence the better estimate of the intercept. For graph (b) in Fig. 6.2 the gradients determined by the two methods are distinct—this is a consequence of the weighted fit giving less importance to the two points with large error bars, which 'pull down' the unweighted line of best fit. We note that the uncertainty in the weighted gradient is also smaller.

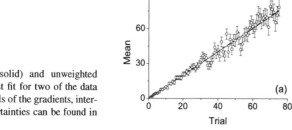

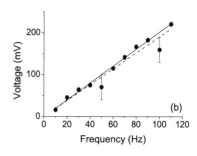

Fig. 6.2 Weighted (solid) and unweighted (dashed) lines of best fit for two of the data sets of Fig. 6.1 Details of the gradients, intercepts and their uncertainties can be found in Table 6.1.

Table 6.1 Details of the gradients and intercepts from Fig. 6.2.

	Graph (a)		Graph (b)	
	Unweighted	Weighted	Unweighted	Weighted
Gradient	1.03 ± 0.02	1.01 ± 0.01	(1.9 ± 0.2) mV/Hz	(2.03 ± 0.05) mV/Hz
Intercept	-0.5 ± 0.9	0.01 ± 0.08	$(0 \pm 1) \times 10$ mV	(-1 ± 3) mV

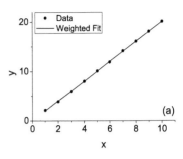

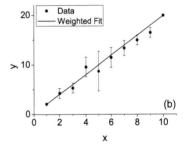

6.3.1 Strategies for a straight-line fit

An interesting feature of Table 6.1, which contains the least-squares best-fit parameters to the data of Fig. 6.2, is the order-of-magnitude reduction in the error bar for the uncertainty with the weighted fit. There are two broad categories of experiments in the physical sciences—the first seeks to verify the form of a physical law as manifest in a mathematical relation; the second seeks to extract a parameter from a well-known physical law. If the experimenter is performing the first type of experiment, there is no *a priori* preference for reducing the error bars for particular data points. In contrast, for experiments for which a linear regression ($y(x_i) = mx_i + c$) will be performed the strategies are clear:

(1) If one wishes to know the intercept with great precision, the best strategy is to invest most of your time and effort in reducing the error bars on the points close to the y-axis. Reducing the error bars on points which are far removed from the intercept will hardly alter the magnitude of the error in the intercept (Fig. 6.1a and b).
(2) If one wishes to know the gradient with great precision, the experimenter should devote most time and effort in reducing the size of the error bars for two points at the extrema of the data set—i.e. the data point with the smallest x-value, and the one with the largest x-value. A more robust measurement will be obtained if the data are recorded over as large a range as can be realised with the experimental apparatus or for which the theory is valid. Again, it is a waste of resource to measure points in the middle of the range of x values as they will hardly influence the error on the gradient (Fig. 6.1c).

These points are further illustrated in Fig. 6.3, which shows four computer-generated heteroscedastic data sets. In parts (a)–(c) the points close to the y-axis have a small error bar, consequently the error in the intercept is small. In part (d), by contrast, the points close to the y-axis have large error bars, leading to a corresponding increase in the error of the intercept. In parts (a) and (b) the error bars of the points at the extrema of the data set are small, hence the error in the slope is small; in parts (c) and (d) for one of the extrema the data are relatively poorly known, which is reflected in the larger uncertainty in the slope.

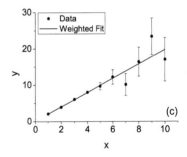

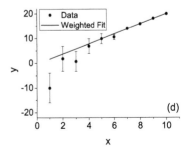

Fig. 6.3 Computer generated data sets with Gaussian noise. For (a), $m = 2.01 \pm 0.01$ and $c = -0.02 \pm 0.07$; (b) $m = 1.99 \pm 0.01$ and $c = 0.04 \pm 0.08$; (c) $m = 1.97 \pm 0.03$ and $c = 0.09 \pm 0.06$ and (d) $m = 2.04 \pm 0.04$ and $c = -0.3 \pm 0.4$.

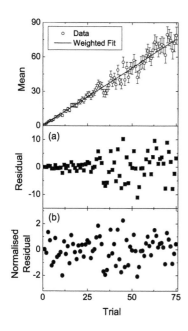

Fig. 6.4 An analysis of the residuals to the fit of graph (a) from Fig. 6.2. The data and the best-fit straight line are shown in the upper panel; part (a) shows the residuals, and panel (b) the normalised residuals. As this is a heteroscedastic data set the raw residuals are difficult to interpret, whereas the normalised residuals are seen to be scattered within ± 2 of zero, as expected for a good fit.

[3] In practice for anything higher than a third-order polynomial numerical techniques are used.

6.3.2   Analysis of residuals with non-uniform error bars

As we saw in Chapter 5 a powerful way to learn rapidly about the quality of the fit is to plot the residuals, $y_i - y(x_i)$. For heteroscedastic data the raw residuals are more difficult to interpret. It is therefore useful to introduce the dimensionless quantity, the **normalised residual**, R_i, defined as

$$R_i = \frac{y_i - y(x_i)}{\alpha_i}, \tag{6.8}$$

where y_i, α_i and $y(x_i)$ are the ith measurement, uncertainty and value of the fit, respectively. An analysis of the residuals to the fit of graph (a) from Fig. 6.2 is shown in Fig. 6.4. The data and the best-fit straight line are shown in the upper panel, part (a) shows the raw residuals, and panel (b) the normalised residuals. We note that the raw residuals grow in magnitude with increasing trial number. However, analysis of the normalised residuals shows that 65% are scattered within ± 1 and 96% are within ± 2, as expected for a good fit. Unlike the case for homoscedastic data, the shape of the histogram of the raw and normalised residuals will be different; the presence of a bias or trend in the data would manifest itself much more obviously in a visual inspection of the normalised residuals. The validity of Gaussian uncertainties of the data is also easier to verify with normalised residuals.

6.4   Performing a weighted least-squares fit—beyond straight lines.

6.4.1   Least-squares fit to an nth-order polynomial

We now consider having a theoretical model which is not restricted simply to a straight line. Consider first the nth-order polynomial:

$$y(x) = a_0 + a_1 x + a_2 x^2 + \cdots + a_n x^n = \sum_{k=0}^{n} a_k x^k. \tag{6.9}$$

Note that the function depends linearly on the coefficients a_k. One can substitute eqn (6.9) into eqn (6.1), and minimise the goodness-of-fit parameter χ^2 by taking partial derivatives with respect to each of the parameters in turn, and setting them to zero. This procedure yields $(n + 1)$ linear coupled equations for the $(n + 1)$ coefficients. It is possible to solve the relevant coupled equations with matrix methods[3] to yield the best-fit coefficients, and their uncertainties (Bevington and Robinson 2003, Section 7.2, Press *et al.* 1992, Section 15.4). It is also possible to use computer software to minimise χ^2, and this is the approach we will use throughout the remainder of the chapter.

6.4.2   Least-squares fit to an arbitrary nonlinear function

Many of the theoretical models which describe experiments performed in the physical sciences involve a nonlinear dependence on some, or all, of the model parameters. In general there will not be an analytic solution for the best-fit parameters. We will still use the goodness-of-fit parameter χ^2 from eqn (6.1) to define the problem, and discuss the numerical methods of obtaining the minimum value of χ^2 in Chapter 7.

The generalisation of eqn (6.9) to an arbitrary nonlinear function with $\mathcal{N}$ parameters is:

$$y(x) = f(x ; a_1, a_2, \ldots, a_{\mathcal{N}}).\qquad(6.10)$$

Using an arbitrary nonlinear function necessitates numerical methods for finding the optimum parameters. The first stage in fitting your data is to define explicitly the theoretical model which describes your data. This choice is often motivated by some physical insight or prior experience.[4] The method for obtaining the best-fit parameters is the following:

[4]It is important to define the problem carefully: a given data set can always be fitted to any function if enough parameters are introduced.

- For each value of the independent variable, x_i, calculate $y(x_i)$ from eqn (6.10) using an estimated set of values for the parameters.
- For each value of the independent variable calculate the square of the normalised residual, $\left[\frac{(y_i - y(x_i))}{\alpha_i}\right]^2$.
- Calculate χ^2 (by summing the square of the normalised residuals over the entire data set).
- Minimise χ^2 by optimising the fit parameters.

The minimum value of χ^2 must be found using numerical methods either by trial-and-error or through more elaborate methods which we discuss in more detail in Chapters 7 and 9. Often a handful of iterations via trial-and-error will yield coarse estimates of the best-fit parameters. Invariably computers are required to facilitate a precise and efficient estimate of the best-fit parameters, especially if the number of parameters or data points is large. As the minimisation procedure of χ^2 is a complex problem in its own right it is important that the minimisation starts with reasonable estimates of the parameters that will be optimised. This ubiquitous method is implemented in data analysis and fitting packages, but can be readily incorporated into either a spreadsheet or fitting algorithm.

We illustrate these ideas by considering the example shown in Fig. 6.5(a) which shows the voltage across the inductor in an *LCR* circuit as a function of time when the circuit is driven externally by a square wave.

A theoretical analysis predicts that the form of the voltage as a function of time is a nonlinear function, $V(t; a_1, a_2, \ldots, a_{\mathcal{N}})$ and we construct a model

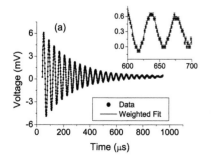

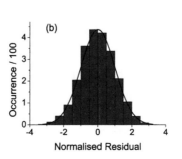

Fig. 6.5 Experimental data of the voltage across an inductor in an *LCR* circuit as a function of time, and the best-fit theoretical model of eqn (6.11). The inset shows detail of the oscillations, and the magnitude of the experimental error bars. The histogram shows the distribution of normalised residuals. Nearly all of the data lie within ± 2 error bars of the theoretical model, which is consistent with a good fit.

which is a damped sinusoidal oscillation with five parameters:

$$V(t) = V_{\text{bgd}} + V_0 \cos\left(2\pi \frac{t}{T} + \phi\right) \exp(-t/\tau). \qquad (6.11)$$

Here, V_{bgd} is a background voltage, V_0 the amplitude of oscillation, T the period of the oscillations, ϕ a relative phase between the driving field and voltage across the inductor, and τ an amplitude-decay, or damping, constant. For this data set, we would like to know the period of oscillations and the damping constant. The data consists of 2200 values of time, voltage and error in voltage (t_i, V_i, α_i). To obtain the best-fit parameters the following procedure was adopted: for each value of t_i, a theoretical voltage $V(t_i)$ was calculated using eqn (6.11) with reasonable estimates of the five parameters, V_{bgd}, V_0, T, ϕ and τ. The estimates of the parameters were deduced by analysing the experimental data. χ^2 was calculated and subsequently minimised by varying all the five parameters to yield the best-fit line shown overlaid with the data in Fig. 6.5(a).

Two obvious questions now arise: (i) is the theoretical model appropriate for the data? and, if so, (ii) what are the errors in the best-fit parameters? We shall address the latter in the next section, and the former in detail in Chapter 8. However we can obtain a qualitative answer to the question 'is the theoretical model appropriate for the data?' by conducting a visual inspection of our fit and the normalised residuals. A good fit will have approximately two-thirds of the data within one error bar of the theoretical function and a histogram of the normalised residuals which is Gaussian. We note that for the particular fit shown in Fig. 6.5(a) approximately two-thirds of the data points are consistent with the theory, and the histogram of the normalised residuals is also well modelled by a Gaussian distribution (the continuous line in Fig. 6.5b).

Given that the model is appropriate, we now show how to answer the question 'what are the uncertainties in the best-fit parameters?' for a nonlinear function. Note that it would be inappropriate to proceed with this analysis if the fit were poor.

6.5   Calculating the errors in a least-squares fit

In Section 5.3.1, Fig. 5.8, we showed how the goodness-of-fit parameter for a simple function, $y = mx$, evolved as m was varied for a given data set. This evolution results in a one-dimensional curve with a clear minimum, which defines the best-fit value. In the more general examples above, where one is fitting more complex functions with non-uniform error bars, the goodness-of-fit parameter remains χ^2, but we now need to consider the evolution of χ^2 over a surface defined by the many fit parameters.[5]

[5]This is often referred to as an $\mathcal{N}$-dimensional hyper-surface.

6.5.1   The error surface

We begin this discussion by considering the special case of a nonlinear function with two parameters, $f(A, B)$, before extending the discussion to the more general case.

For a two-parameter function the equivalent of Fig. 5.8 in Section 5.3.1 is now a two-dimensional *surface* which is shown schematically in Fig. 6.6. The coordinates of the point in the $A-B$ plane at which χ^2 is minimum define the best-fit values of the fit parameters, $\overline{A}$ and $\overline{B}$, shown by the dot in the centre. In the plot we also show contours of constant χ^2 around the minimum value, χ^2_{min}. The shape of the **error surface** in the vicinity of the minimum shows how the fit is sensitive to the variations in the fit parameters. A high density of contours along a given parameter axis indicates high sensitivity to that parameter and conversely a sparse contour density shows that the fit is insensitive to that parameter. In general the shape of a fixed χ^2 contour will be elliptical. The tilt of the contours yields information about the correlation between the uncertainties in parameters, a topic which is discussed extensively in Chapter 7. When the ellipse axes coincide with the parameter axes, the uncertainties in the parameters are independent. For the remainder of this chapter we will assume that the contour plot of χ^2 is a well-behaved function with a single minimum. Strategies for dealing with more complex error surfaces will be outlined briefly in Chapter 9.

One can investigate the shape of the error surface in the vicinity of the minimum by performing a Taylor-series expansion of χ^2. For one parameter the second order of expansion reads:

$$\chi^2\left(\overline{a}_j + \Delta a_j\right) = \chi^2\left(\overline{a}_j\right) + \frac{1}{2}\frac{\partial^2 \chi^2}{\partial a_j^2}\bigg|_{\overline{a}_j}\left(\Delta a_j\right)^2, \tag{6.12}$$

where Δa_j is the excursion of the parameter away from its optimal value, $\overline{a}_j$. There is no linear term in eqn (6.12) as the first-order derivative of χ^2 disappears at the minimum. In the next section we will see that for an excursion equal in magnitude to the size of the error bar, the value of χ^2 will increase by 1. This allows us to write the standard error in the parameter in terms of the **curvature** of the error surface:

$$\alpha_j = \sqrt{\frac{2}{\left(\frac{\partial^2 \chi^2}{\partial a_j^2}\right)}}. \tag{6.13}$$

We see that a large curvature of χ^2 near the optimal value results in a small standard deviation of the parameter a_j. This is a mathematical way of relating the contour density to the uncertainty in the parameter. We will return to this mathematical formalism for evaluating the uncertainties in parameters from the curvature of the error surface in Chapter 7, where we will discuss the correlation of the parameters, and thus the correlations between the parameter uncertainties, in terms of the Hessian matrix.

6.5.2 Confidence limits on parameters from weighted least-squares fit

Having ascertained the best-fit parameters from a weighted least-squares fit and undertaken a qualitative analysis of the quality of the fit, the question

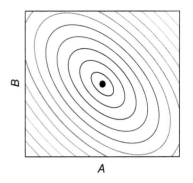

Fig. 6.6 Contours of constant χ^2 in the $A-B$ plane for a general two-parameter nonlinear function $f(A, B)$. The minimum value of χ^2, χ^2_{min}, is obtained with the best-fit values of the parameters, $\overline{A}$ and $\overline{B}$, shown by the dot in the centre. The contours increase in magnitude as one departs from the best-fit parameters.

Table 6.2 Confidence limits associated with various $\Delta\chi^2$ contours for one degree of freedom.

$\Delta\chi^2$ contour	1.00	2.71	4.00	6.63	9.00
Measurements within range	68.3%	90.0%	95.4%	99.0%	99.7%
	1σ		2σ		3σ

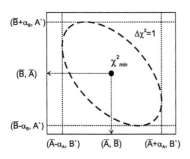

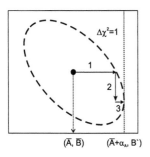

Fig. 6.7 The $\Delta\chi^2 = 1$ contour in the A–B plane. The horizontal and vertical tangent planes define the uncertainties in the parameters.

Fig. 6.8 The procedure for obtaining the uncertainties in the best-fit parameters. χ^2_{min} is achieved when the parameters have the values $\overline{A}, \overline{B}$. If A is increased with B held constant one follows the trajectory labelled 1 to arrive at the desired $\Delta\chi^2$ contour. Then, B must be allowed to vary to reduce χ^2, which results in motion across the error surface along the path labelled 2. This iterative procedure is repeated until an extremum of the desired $\Delta\chi^2$ contour is achieved, where the coordinate of the abscissa yields $\overline{A} + \alpha_A$. Repeating this procedure from $\overline{A}, \overline{B}$ by keeping A constant and increasing B will give the error bar for B.

remains 'what are the errors on the best-fit parameters?'. Specifically, one would like to know the 68% (1σ), 95.4% (2σ), etc. confidence limits. We have seen that the value of χ^2 is an indicator of the goodness of fit and in the previous section how the variation of χ^2 around its minimum is related to the sensitivity of the fit to the individual parameters. We can therefore use the variation in χ^2 from its minimum value, χ^2_{min}, which is $\Delta\chi^2$, to characterise the uncertainties in the best-fit parameters.

How do we determine which $\Delta\chi^2$ contour represents the 68% confidence limit for a given parameter? The probability of χ^2 changing by a certain amount is given by the probability distribution function of χ^2 (discussed in detail in Chapter 8) with the appropriate degrees of freedom. From the PDF of χ^2 for one degree of freedom we can calculate the $\Delta\chi^2$ values that correspond to a particular confidence limit. For one degree of freedom these $\Delta\chi^2$ contours are given in Table 6.2.

In the example shown in Fig. 6.7, the $\Delta\chi^2 = 1$ contour is plotted. This is the contour that corresponds to the confidence limit that we have adopted for our error bar throughout this book. This contour defines a confidence region in the A–B plane for both parameters. However, what we require is the confidence limit for a single parameter—in this case either A or B. To extract the confidence limit in one of the parameters, we do not read off the intersection of the relevant contour with the parameter axis, but rather take the extremum of the ellipse. As shown in Fig. 6.7 there are four extrema of the ellipse; two are extrema for B and occur at the horizontal tangent planes, with the other two being extrema for A occurring at the vertical tangent planes. The difference between the value of A at which the extremum occurs and the best-fit value, $\overline{A}$, defines the error bar, α_A.

To arrive at the extremum of a particular $\Delta\chi^2$ contour requires a series of orthogonal steps across the error surface. So, if one wishes to find the projection of a particular contour in A, the first stage would be to move along the error surface from the location of χ^2_{min} changing A from its optimal value $\overline{A}$, until the appropriate contour is reached (path 1 in Fig. 6.8). Having reached this point, an orthogonal motion is undertaken by re-optimising B whilst keeping A at its new value (path 2 in Fig. 6.8). These two steps are repeated until no discernible change is observed in the χ^2 value. It is not too difficult to automate this process in most spreadsheet packages. The difference between the final iterated value of A, $\overline{A} + \alpha_A$, and the optimal value, $\overline{A}$, is the uncertainty in A. The mathematical operations required to move to the extremum of a particular $\Delta\chi^2$ contour are two independent χ^2 minimisations. For the special case of a two-parameter fit, both operations are χ^2 minimisations with one degree of freedom.

Extending the procedure for $\mathcal{N}$ parameters $(a_1, a_2, \ldots, a_{\mathcal{N}})$ is similar to that described above for the two-dimensional case. The error surface becomes $\mathcal{N}$-dimensional. Navigating this surface involves making a change to one of the parameters from its optimal value as before and then re-optimising all other $\mathcal{N} - 1$ parameters. As before, both operations are independent χ^2 min-imisations. The minimisation where the parameter of interests is changed has one degree of freedom; the re-optimisation of all other parameters has many degrees of freedom. Thus, for the uncertainty in the single variable of interest, the $\Delta\chi^2$ contours which correspond to a particular confidence level remain those listed in Table 6.2. If one is interested in the other confidence limits listed in Table 6.2 one follows the same procedure to locate the extremum of the appropriate $\Delta\chi^2$ contour.

A summary of calculating errors on parameters using χ^2 minimisation

(1) Find the best-fit parameters by minimising χ^2.
(2) Check the quality of the fit. If the fit is poor, there is little point in using, or calculating, the errors in the best-fit parameters.
(3) Adjust the parameter whose error you wish to determine until χ^2 becomes $\chi^2_{\text{min}} + 1$.
(4) Re-minimise χ^2 using all the other parameters, but *not* the parameter whose error you are measuring.
(5) Iterate 3 and 4 until the value of χ^2 does not change (within some tol-erance). The standard error on the parameter is the absolute difference between the current and optimal value of that parameter.
(6) Repeat steps 3–5 for each of the remaining fit parameters of interest.

We now illustrate some of these abstract concepts explicitly through worked examples.

6.5.3 Worked example 1—a two-parameter fit

Figure 6.9 shows the $\Delta\chi^2$ contours for the two parameters (gradient, m, and intercept, c) of the weighted least-squares straight-line fit to the data of Fig. 6.1(d). The coordinates of the location of χ^2_{min} are $\overline{m} = 2.03$ mV/Hz, and $\overline{c} = -1$ mV, in agreement with Table 6.1. Some of the contours of Table 6.2 are indicated on Fig. 6.9.

We can obtain the error in the slope by looking at the coordinates of the extremum of the $\Delta\chi^2 = 1$ contour along the gradient axis. The lower right-hand extremum for m is located at $m = 2.08$ mV/Hz, giving the uncertainty in m as $\alpha_m = 2.08 - 2.03 = 0.05$ mV/Hz. A similar procedure using the lower extremum for the intercept gives $\alpha_c = -1 - (-4) = 3$ mV.[6] We obtain iden-tical values of the error for either parameter if we choose the other extremum.

The fact that the axes of the contours of χ^2 around the minimum are not coincident with the m–c axes is an indication of the degree of correlation between the variables. We will discuss the implications of such a correla-tion in Chapter 7. The importance of allowing the other parameters to be

[6] The extremum of the $\Delta\chi^2 = 1$ contour along one axis *does not* occur at the same coordinates as the extremum along the orthogonal axis. Thus two separate iterations are required to obtain the value of the errors for both parameters.

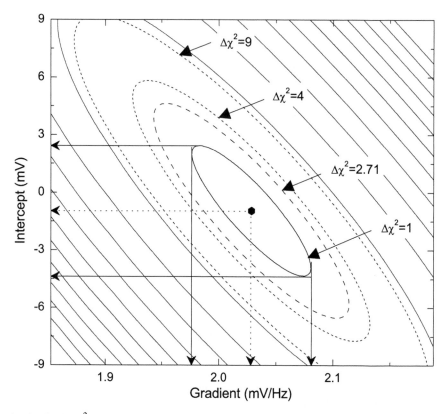

Fig. 6.9 A contour plot showing iso-χ^2 lines for the gradient and intercept of the straight-line fit to the data of Fig. 6.1(d). The $\Delta\chi^2$=1, 4, 9 and 2.71 contours correspond, respectively, to one, two, three standard deviations, and the 90% confidence limit.

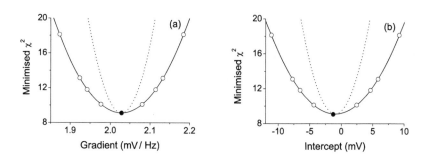

Fig. 6.10 Cuts through the error surface of Fig. 6.9 as a function of (a) the gradient, and (b) the intercept. The dashed lines are obtained when one parameter is varied, and the other held at its optimum value; the solid line is obtained when one parameter is varied, and the other re-optimised to minimise χ^2. The open circles correspond to increases in χ from χ^2_{min} of 1, 2.71, 4 and 9 and are the extrema along the appropriate axes of the ellipses in Fig. 6.9.

re-optimised when evaluating the error in one of the parameters is highlighted in Fig. 6.10.

Parts (a) and (b) of Fig. 6.10 show the variation of χ^2 with respect to the gradient and intercept, respectively. The dashed curves show the variation with respect to one parameter, when the other is left at its optimum value. In contrast, the solid curves show the variation with respect to one parameter, when the other is re-optimised. It is the latter which yields the confidence limit on the parameters. The $\Delta\chi^2 = 1$ contour is reached at a gradient of 2.05 mV/Hz if m is increased and the intercept kept as $\bar{c}$ which gives an unrealistic error of 0.02 mV/Hz. However, following the procedure depicted in Fig. 6.8, the extremum is at a value of 2.08 mV/Hz and the uncertainty is 0.05 mV/Hz, more than double the value obtained without re-optimising the intercept. The more tilted the contours are, the more extreme this difference becomes.

6.5.4 Worked example 2—a multi-parameter fit

Here we look at an example of a four-parameter fit. The data in Fig. 6.11(a) show the number of counts per second (c.p.s.) when X-rays are diffracted from the (111) peak of a thin film of permalloy as a function of the scattering angle. The error bars come from Poisson count statistics. The model to describe the data has four parameters: the height of the Lorentzian lineshape, S_0, the angle at which the peak is centred, θ_0, the angular width of the peak, $\Delta\theta$, and a constant background offset, S_{bgd}. Mathematically, the signal, S, is of the form:

$$S = S_{bgd} + \frac{S_0}{1 + 4\left(\frac{\theta - \theta_0}{\Delta\theta}\right)^2}. \tag{6.14}$$

The error surface for this problem is four dimensional. To extract the uncertainty in a parameter one needs to reduce this four-dimensional surface to a single one-dimensional slice as shown in Fig. 6.11(d). This is achieved by varying the parameter of interest and allowing all other parameters to be re-optimised to minimise χ^2. Additionally it is useful to consider two-dimensional slices of the error surface to investigate the correlation of the parameters. These 2D surfaces are created by varying two of the parameters and allowing all other parameters to be re-optimised to minimise χ^2. In Fig. 6.11(b) and (c) we show the two-dimensional surface contours of $\Delta\chi^2$ where we investigate the correlation between the background and peak centre in (b) and the background with peak width in (c). In Fig. 6.11(b) the contour ellipses close to χ^2_{min} are aligned with the coordinate axes, indicating that there is very little correlation between the background and the peak centre. In contrast the contours are tilted with respect to the axes in Fig. 6.11(c), indicating a negative correlation between background and peak width. The rotation of the ellipse is a manifestation of the fact that the background and width can be played off against each other to a certain extent—an increase of the background can be partially compensated for by a decrease of the width of the peak. The influence on the uncertainty of one parameter owing to correlation with other parameters is taken into account by allowing all the other parameters to be re-optimised when calculating the one-dimensional variation of χ^2 with respect to that parameter in the vicinity of the minimum (Fig. 6.11d).

6.6 Fitting with constraints

A physical insight often allows the number of independent fitting parameters to be reduced. A constraint is easily incorporated into most fitting packages or a spreadsheet.

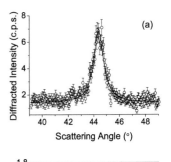

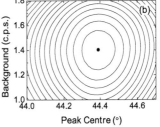

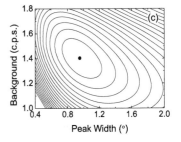

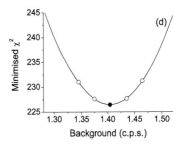

Fig. 6.11 (a) An X-ray diffraction peak as a function of scattering angle. The model comprises a Lorentzian peak on a constant background. (b) Contour plot of the background and centre of the peak. (c) Contour plot of peak width and background. (d) The variation of χ^2 around the optimum background value, where the other three parameters are re-optimised for each background value.

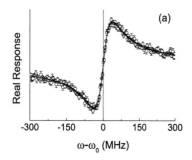

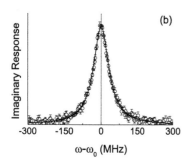

Fig. 6.12 Plots of (a) the real and (b) the imaginary parts of the electrical susceptibility of a sample of laser-cooled atoms. The former is proportional to the refractive index of the medium, the latter to the absorption coefficient. These quantities are related via the Kramers–Kronig relations.

Consider the data shown in Fig. 6.12 displaying the (a) real and (b) imaginary parts of the electrical susceptibility of a medium comprising laser-cooled atoms. The data in part (a) follow the dispersion curve associated with the spectral dependence of the refractive index, whereas the absorption spectrum of (b) displays a Lorentzian lineshape. The Kramers–Kronig relations explain how the knowledge of the spectral response of either the real or imaginary part of the susceptibility enables the other to be calculated. The theoretical prediction is that the complex electrical susceptibility, χ_E, in this situation is

$$\chi_E = \chi_0 \frac{\Delta}{\Delta - i\Gamma}, \tag{6.15}$$

where $\Delta = \omega - \omega_0$ is the detuning (the difference between the applied frequency and the resonant frequency), and Γ is the linewidth. By calculating eqn (6.15) in a spreadsheet, with two parameters, the real and imaginary parts can be compared with the experimental data and their uncertainties. A least-squares fit can then be conducted simultaneously on both data sets. Simultaneous data fitting is difficult to achieve using a fitting package, but is relatively straightforward in a spreadsheet. The thick lines on the figures show the results of such a minimisation procedure. The result obtained for the linewidth with this procedure is $\Gamma = 37.69 \pm 0.03$ MHz; when the two data sets are analysed individually, less precise results are obtained: $\Gamma = 37.71 \pm 0.04$ MHz from the real part, and $\Gamma = 37.66 \pm 0.05$ MHz from the imaginary component.

During a minimisation procedure it is possible, on occasion, for χ^2 to be reduced by extending the value of a parameter outside a range for which it is physically realistic. Most fitting or spreadsheet packages enable constraints to be applied during the minimisation. The experimenter should ensure that after minimisation all parameters remain within a physically realistic domain, even if this means a higher χ^2_{min} value.

Sometimes the experimenter has some prior knowledge which can be useful in constraining fit parameters. Examples include branching ratios, the ratio of masses, mass differences, and atomic energy level intervals. Figure 6.13(a) shows an X-ray diffraction spectrum of the (3 1 1) peak of Cu. There are, in fact, two unresolved peaks which have a width greater than their angular separation. It is known that the peaks must have the same lineshape function, the same angular width, and a well-defined angular separation. A multi-dimensional χ^2 fit was performed, and the result is shown as the solid line in Fig. 6.13. In this case the relative heights of the peaks was allowed to float as a free parameter, but analysis of the optimal parameters gave a ratio of the intensity of the two peaks as 2.1 ± 0.1, consistent with the theoretical value of 1.96. One could have used the theoretical value to constrain the parameters (i.e. reduce the number of fitting parameters by 1), or, as here have allowed them to float. Although this increases the number of fit parameters it is a useful test of the validity of the theoretical model, especially when the number of data points greatly exceeds the number of parameters. Note that allowing the widths of the two components to vary independently, in this case, leads to inconsistent results with poor precision; similarly, fitting the data with a single-peak function results in a poorer fit with unrealistic peak widths.

6.7 Testing the fit using the residuals

In Chapter 5 and in section 6.3.2 we stressed the importance of investigating the normalised residuals as an indicator of the quality of a fit. We return to this topic in this section and show how an incorrect model can lead to structure and autocorrelations in the normalised residuals. In Fig. 6.13 we show two fits to the X-ray diffraction spectrum from the (3 1 1) peak of Cu that was discussed in Section 6.6. In Fig. 6.13(a) the correct constrained model with two peaks is used to fit the data, whereas in (b) we show the fit to an erroneous model based on a single peak. A quick visual inspection of the two figures shows that both fits reproduce most of the main features, but a closer inspection of the residuals shows structure in the normalised residual plot close to the fitted peak centre in Fig. 6.13(b).

Structure in the residuals can be visualised more clearly by making use of a **lag plot** which is constructed by plotting the normalised residual R_i (see eqn 6.8) against a lagged residual R_{i-k} where k is an integer, and is usually 1. These plots are useful for identifying outliers and for testing the randomness of the errors. Any non-random pattern in the lag plot is an indicator of autocorrelations in the residuals and suggests something is missing from the theoretical model. In Fig. 6.14 the lag plots for $k = 1$ are shown for the two fits in Fig. 6.13. The lag plot for the constrained fit (Fig. 6.14a) shows no structure, but that of the single-peak fit (Fig. 6.14b) clearly shows a positive trend suggesting a positive correlation between some of the lagged residuals. For a good fit we expect that, if the uncertainties in the data are Gaussian, 91% of the normalised residuals should fall within a two-dimensional box defined by the ± 2 limits on the lag plot. For the constrained fit, the number of residuals inside this box is 93%, in excellent agreement with this hypothesis. On the other hand, the single-peak fit only has 82% of the data points which are within the box, indicative of a poor fit.

The degree of correlation, or shape, in the lag plot can be reduced to a single numerical value by evaluating the *Durbin–Watson* (Durbin and Watson 1950) statistic, $\mathcal{D}$. In its weighted form, for $k = 1$, $\mathcal{D}$ is defined in terms of the normalised residuals:

$$\mathcal{D} = \frac{\sum\limits_{i=2}^{N} [R_i - R_{i-1}]^2}{\sum\limits_{i=1}^{N} [R_i]^2}. \tag{6.16}$$

The Durbin–Watson statistic has values in the range $0 < \mathcal{D} < 4$ and may indicate three limiting cases:

(1) $\mathcal{D} = 0$: systematically correlated residuals;
(2) $\mathcal{D} = 2$: randomly distributed residuals that follow a Gaussian distribution;
(3) $\mathcal{D} = 4$: systematically anticorrelated residuals.

An experimenter should be highly suspicious if the value of $\mathcal{D}$ approaches either 0 or 4, and question the fit should $\mathcal{D}$ differ significantly from 2. For

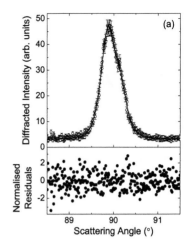

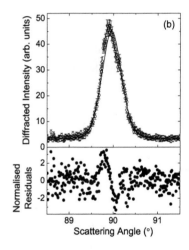

Fig. 6.13 Experimental spectra and weighted least-squares-fit (thick line) for the Cu (3 1 1) X-ray diffraction peak. In (a) the fit and residuals are shown for a constrained double peak model and in (b) the experimental data are fitted to a single-peak function. The residuals are randomly distributed for the double-peak fit, but show structure for the single-peak fit.

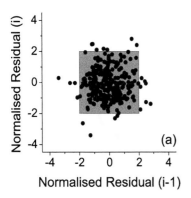

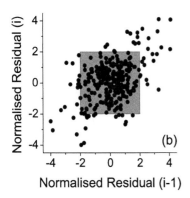

Fig. 6.14 Lag plots with $k = 1$ of the normalised residuals shown in Fig. 6.13. (a) The lag plot from the two-peak fit with 93% of the normalised residuals within $\pm 2\sigma$ of the zero mean. (b) The lag plot from Fig. 6.13(b) in which a clear linear trend is observed and only 82% of the data are within the expected confidence limit. The Durbin–Watson statistic $\mathcal{D}$ is 1.97 for fit (a) and 1.12 for (b).

homoscedastic data sets it is possible to use the $\mathcal{D}$ statistic to test the quality of the fit but in Chapter 8 we will discuss a more robust test based on the χ^2 statistic. For the examples shown in Fig. 6.13 the Durbin–Watson statistic is 1.97 for the constrained fit and 1.12 for the erroneous fit. As $\mathcal{D}$ is very close to 2 for the constrained double-peak fit we would expect that this model was better than the alternative single peak for which $\mathcal{D}$ was significantly different from 2.

Chapter summary

- A sequence of data points with non-uniform error bars is referred to as **heteroscedastic**.
- The goodness-of-fit parameter is $\chi^2 = \sum_i \dfrac{(y_i - y(x_i))^2}{\alpha_i^2}$, with $y(x_i)$ the theoretical model, y_i the data and α_i the corresponding error bars.
- For linear regression, $y(x_i) = m\,x_i + c$, with non-uniform error bars there exist analytic solutions for the best values of the slope, m, the intercept c and their associated errors:

$$m = \frac{\sum_i w_i \sum_i w_i x_i y_i - \sum_i w_i x_i \sum_i w_i y_i}{\sum_i w_i \sum_i w_i x_i^2 - \left(\sum_i w_i x_i\right)^2},$$

$$\alpha_m = \sqrt{\frac{\sum_i w_i}{\sum_i w_i \sum_i w_i x_i^2 - \left(\sum_i w_i x_i\right)^2}},$$

$$c = \frac{\sum_i w_i x_i^2 \sum_i w_i y_i - \sum_i w_i x_i \sum_i w_i x_i y_i}{\sum_i w_i \sum_i w_i x_i^2 - \left(\sum_i w_i x_i\right)^2},$$

$$\alpha_c = \sqrt{\frac{\sum_i w_i x_i^2}{\sum_i w_i \sum_i w_i x_i^2 - \left(\sum_i w_i x_i\right)^2}},$$

where the weighting for each point is the inverse square of the uncertainty, $w_i = \alpha_i^{-2}$, and the summation is over all the data points.
- The best-fit parameters for a nonlinear function are usually found by computer minimisation. The best-fit parameters are the ones for which χ^2 is at a minimum, χ^2_{min}. The uncertainty in a given parameter can be found by an iterative procedure: the parameter A is increased from the best value, $\bar{A}$, with all the other parameters held constant until χ^2 has increased by 1. Then, all the other parameters must be allowed to vary to reduce χ^2. A is increased again, and the process repeated until convergence is obtained. The value of A at which this occurs is $\bar{A} + \alpha_A$.

Exercises

(6.1) *Linear regression—unweighted fit*
The data plotted in Fig. 6.1(d) are listed below.

Frequency (Hz)	10	20	30	40	50	60
Voltage (mV)	16	45	64	75	70	115
Error (mV)	5	5	5	5	30	5

Frequency (Hz)	70	80	90	100	110
Voltage (mV)	142	167	183	160	221
Error (mV)	5	5	5	30	5

Use the results of Chapter 5 to ascertain the best-fit gradient and intercept using an unweighted fit. Verify that the results are in agreement with Table 6.1.

(6.2) *Linear regression—weighted fit*
For the data set of Exercise (6.1) use the results of eqns (6.3)–(6.6) to calculate the best-fit gradient and intercept using a weighted fit. Verify that the results are in agreement with Table 6.1. Draw the lag plot, and calculate the Durbin–Watson statistic $\mathcal{D}$.

(6.3) *Normalised residuals*
Use the fit parameters from the last question to plot the best-fit straight line with the data. Use eqn (6.8) to calculate the normalised residuals, and plot them.

(6.4) *Error bars from a χ^2 minimisation*
(i) For the data set of Exercise (6.1), write a spreadsheet which you can use to perform a χ^2 minimisation. Verify that χ^2_{min} is obtained for the same values of the parameters as are listed in Table 6.1. (ii) By following the procedure of $\chi^2 \rightarrow \chi^2_{min} + 1$ outlined in Section 6.5, check that the error bars for m and c are in agreement with Table 6.1. Let the parameters (a) *increase* from their optimal values to find the extremum for each parameter, and (b) *decrease* from their optimal values to find the extremum for each parameter. Are the uncertainties determined in (a) and (b) the same? (iii) Calculate the uncertainties on m and c by finding the extrema of the $\chi^2 \rightarrow \chi^2_{min} + 4$ and $\chi^2 \rightarrow \chi^2_{min} + 9$ contours. Are your results in agreement with Table 6.2?

(6.5) *Speed of light*
The data plotted in Fig. 6.1(c) are listed below.

Displacement (m)	0.05	0.25	0.45	0.65	0.85
Phase (rad)	0.00	0.21	0.44	0.67	0.88
Error (rad)	0.05	0.05	0.05	0.05	0.09

Displacement (m)	1.05	1.25	1.45	1.65	1.85
Phase (rad)	1.1	1.3	1.5	2.0	2.24
Error (rad)	0.1	0.2	0.5	0.1	0.07

The speed of light is related to the slope, m, of this graph via the relationship speed of light $= \frac{2\pi \times 60 \times 10^6 \mathrm{Hz}}{m}$. Calculate the the slope and intercept of the best-fit straight line to the data, and their associated errors. Hence deduce the speed of light, and its error. The theoretical predication is that the intercept should be zero. Is this consistent with the data?

(6.6) *Strategies for error bars*
Consider the following data set:

x	1	2	3	4	5
y	51	103	150	199	251
α_y	1	1	2	2	3

x	6	7	8	9	10
y	303	347	398	452	512
α_y	3	4	5	6	7

(i) Calculate the weighted best-fit values of the slope, intercept and their uncertainties. (ii) If the data set had been homoscedastic, with all the errors equal, $\alpha_y = 4$, calculate the weighted best-fit values of the slope, intercept and their uncertainties. (iii) If the experimenter took greater time to collect the first and last data points, for which $\alpha_y = 1$, at the expense of all of the other data points, for which $\alpha_y = 8$, calculate the weighted best-fit values of the slope, intercept and their uncertainties, and comment on your results.

Computer minimisation and the error matrix

7

As we discussed in Chapter 6, the errors on the parameters, α_i, are related to how the χ^2 hyper-surface varies around the minimum value, $\chi^2_{\min}$. The local curvature of the error hyper-surface is mapped out by calculating χ^2 as the parameters are displaced from their optimal values. Therefore, it is not a surprise that there is a mathematical relationship between the curvature of the hyper-surface and the magnitude of the errors of the fitted parameters, see eqn (6.13). In this chapter we will develop a general matrix methodology for describing the variation of the χ^2 surface and for determining the uncertainties in the fit parameters. We begin by discussing different types of algorithms used in fitting packages to minimise χ^2; this both serves to reinforce the discussion of finding the best fit described in earlier chapters, and is a convenient way of introducing the matrices from which the uncertainties in the fit parameters can be deduced.

In the previous chapters we have implicitly assumed that both our parameters and their errors are independent. However, the statistical fluctuations of the data lead to correlations among the uncertainties in the best-fit values of the independent parameters. In Section 7.3 we detail how this degree of correlation can be quantified and incorporated into error analysis.

The data are fitted using the standard chi-squared statistic, χ^2, we have seen previously. In general, we wish to fit our N data points, (x_i, y_i), with uncertainties given by α_i, to a nonlinear function with $\mathcal{N}$ parameters, $f(x; a_1, a_2, \ldots, a_\mathcal{N})$. The analytic results of least-squares fitting discussed in earlier chapters for linear regression cannot be used for nonlinear functions, consequently we develop approximate iterative procedures for locating the minimum of the multi-dimensional error surface.

7.1 How do fitting programs minimise?

We now return to discussing how computer fitting functions arrive at the best fit. Irrespective of whether the minimisation process occurs within a spreadsheet or a fitting program, all minimisation techniques are based on an iterative method of improving a trial solution by a reduction in the goodness-of-fit parameter. There are several different approaches that can be adopted to navigate over the $\mathcal{N}$-dimensional error surface to arrive at the minimum. In this chapter we restrict our discussion to the easiest case where there is a single minimum on the error surface. We outline five possible minimisation

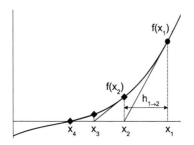

Fig. 7.1 The Newton–Raphson method for finding the zero crossing of a function. After s iterations the function is $f(x_s)$, and its derivative at x_s is $f'(x_s)$. The tangent to the curve is drawn at x_s, and the zero crossing of the tangent found. This point is taken as the updated coordinate of the zero-crossing point of the function, x_{s+1}. The iterative procedure continues until convergence at the desired tolerance.

methods here; further details for these, and alternative, strategies can be found in references such as Bevington and Robinson (2003, Chapter 8) and Press *et al.* (1992, Chapters 10 and 15). Each of the techniques discussed in detail are **iterative**, and proceed by applying an update rule. We first discuss these concepts in the more familiar context of finding the zero of a function.

7.1.1   Iterative approaches

The Newton–Raphson method is a numerical technique to solve the equation $f(x) = 0$. It is an iterative procedure illustrated in Fig. 7.1. We assume that the first approximate solution is x_1, where the function is $f(x_1)$, and the first derivative $f'(x_1)$. Let the zero crossing of $f(x)$ be at $x_1 + h$, then $f(x_1 + h) = 0$. From Fig. 7.1 we see that if h is small

$$f(x_1 + h) \approx f(x_1) + f'(x_1)h,$$

$$\therefore f(x_1) + f'(x_1)h \approx 0,$$

$$\therefore h \approx -\frac{f(x_1)}{f'(x_1)}. \tag{7.1}$$

This allows us to write down a second approximation to the zero crossing:

$$x_2 = x_1 - \frac{f(x_1)}{f'(x_1)}. \tag{7.2}$$

The process can be repeated to obtain successively closer approximations. If after s iterations the approximate solution is x_s then the next iteration is x_{s+1}, and these quantities are related via the relation:

$$x_{s+1} = x_s - \frac{f(x_s)}{f'(x_s)}. \tag{7.3}$$

Equation (7.3) is known as the **update rule**, as it allows the parameter (the current best value of the zero crossing) to be amended. The procedure continues until the zero-crossing point is found, to within a certain tolerance. If the first guess for the solution is close to the actual solution, the procedure converges rapidly. Conversely, if the first guess is far from the optimal value there are two issues: (i) the convergence can be slower; and (ii) depending on the sign of the gradient of the function it is possible that the updated solution will be further from the optimal value than the current value. Therefore it is important to have a reasonable first guess as the input to utilise the power of the Newton–Raphson method. The convergence properties of this method are discussed in Exercise (7.1), and the procedure can easily be extended to find maxima or minima of a function, as shown in Exercise (7.2).

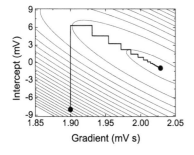

Fig. 7.2 An illustration of a grid search on a two-dimensional error surface. Initial trial values of the gradient and intercept were chosen (gradient 1.90 mV s, intercept -8 mV), and each parameter was optimised in turn until convergence and the minimum χ^2 was obtained. Note the inefficient zig-zag motion along the flat valley bottom.

7.1.2   Grid search

We illustrate the grid search method for finding the minimum of the error surface by applying it to the case of Fig. 6.1(d) (the results of the minimisation procedure are summarised in Table 6.1), as seen in Fig. 7.2. Starting values of the trial parameters (gradient 1.90 mV s, intercept -8 mV) and step sizes are chosen. The minimisation proceeds through an iterative process whereby the goodness-of-fit parameter is minimised by changing each parameter in turn. From the initial position the gradient is kept the same, with the intercept increased. If the value of χ^2 increases, the direction of motion across the error surface is reversed. Having started on a downhill trajectory, the values of the parameter are changed until χ^2 increases. Using the last three increments of the parameter, it is easy to find the value at which the minimum occurs, see Exercise (7.3)—this procedure is known as a **line minimisation**. The procedure described above is then repeated for the intercept as shown in Fig. 7.2. This iterative procedure converges toward the local minimum. The grid-search method is powerful in that it is easy to implement computationally if fitting to a model which has analytic derivatives.

For a model with more than two parameters a similar procedure is applied, with each parameter minimised in turn. The whole procedure is then repeated until the **convergence tolerance** is met—the procedure is terminated when the value of χ^2 changes by less than, say, one part in 10^3. If the variation of χ^2 with respect to each parameter is insensitive to the values of the other parameters, the axes of the χ^2 contours are coincident with the parameter axes; under these conditions the grid-search method converges rapidly. In the more general case, as depicted in Fig. 7.2, there is a degree of correlation among the parameters, and the contours of constant χ^2 are tilted. Figure 7.3 shows the evolution of χ^2 and the parameters as functions of the number of iterations. The trajectory across the error surface zig-zags, which is very inefficient, and the convergence is slow. The advantages of the grid-search method are (i) simplicity, and (ii) that the minimisation process is relatively insensitive to the initial trial parameters. The major disadvantage is that convergence to the minimum can be very slow for correlated parameters (the more general case).

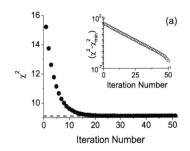

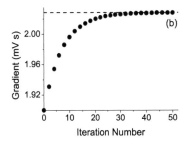

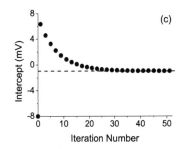

Fig. 7.3 The evolution of χ^2, and the parameters as a function of iteration number for the grid-search method. $\chi^2 - \chi^2_{\min}$ is shown on a log scale in the inset of (a).

7.1.3 The gradient-descent method

In the grid-search method each parameter was varied sequentially which is not efficient. A more elegant approach is to increment all the parameters simultaneously, with the aim of having a vector directed towards the minimum. The vector $\nabla\chi^2$ lies along the direction along which χ^2 varies most rapidly on the error surface. The minimisation proceeds by taking steps along this **steepest descent**. The gradient along the direction of the parameter a_j can be approximated numerically by the formula

$$\left(\nabla\chi^2\right)_j = \frac{\partial\chi^2}{\partial a_j} \approx \frac{\chi^2\left(a_j + \delta a_j\right) - \chi^2\left(a_j\right)}{\delta a_j}, \tag{7.4}$$

where δa_j is a small increment in a_j (typically much smaller than the step size). Note that it is possible within the framework of some minimisation algorithms to enter analytic formulae for the gradient with respect to the parameters.

Let the $\mathcal{N}$-dimensional column vector $\mathbf{a}_s$ contain the $\mathcal{N}$ parameters a_j after s iterations. The update rule is

$$\mathbf{a}_{s+1} = \mathbf{a}_s - \beta\,\nabla\chi^2\left(\mathbf{a}_s\right), \tag{7.5}$$

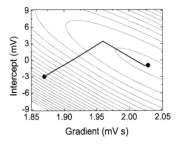

Fig. 7.4 An illustration of a gradient-descent method on a two-dimensional error surface. Each parameter is optimised simultaneously along a vector defined using the gradient.

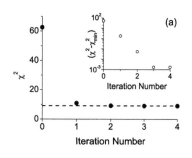

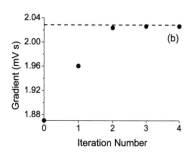

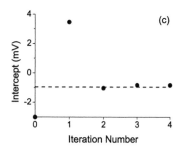

Fig. 7.5 The evolution of χ^2 and the parameters as a function of iteration number for the gradient-descent method. $\chi^2 - \chi^2_{min}$ is shown on a log scale in the inset of (a).

[1] This is a matrix version of eqn (6.12) generalised to $\mathcal{N}$ dimensions; written explicitly as a summation over components it would read

$$\chi^2 (a_s + h) \approx \chi^2 (a_s) + \sum_{j=1}^{\mathcal{N}} \frac{\partial \chi^2}{\partial a_j}\bigg|_{a_s} \Delta a_j$$

$$+ \frac{1}{2} \sum_{j=1}^{\mathcal{N}} \frac{\partial^2 \chi^2}{\partial a_j^2}\bigg|_{a_s} (\Delta a_j)^2 .$$

where the value of the scaling factor β is chosen such that the trajectory proceeds downhill, but does not overshoot up the ascending side of a valley, see Fig. 7.4. After each step, the direction of the steepest descent is calculated again and the process iterated. The method of steepest descent is good at locating an approximate minimum, but is typically augmented by other methods to locate the actual minimum. Major problems with this approach are that as the goodness-of-fit parameters approach the minimum (i) the direction chosen by the method of steepest descent is not the most efficient, and (ii) the numerical determination of the derivative (eqn 7.4), which involves the subtraction of two similar numbers, is error prone. For the case of Fig. 6.1(d) the gradient-descent method converges to χ^2_{min} at the 10^{-3} level within three iterations, see the inset to Fig. 7.5, whereas the grid-search method takes nearly 30 iterations to achieve the same level of convergence as seen in the inset to Fig. 7.3. Note, however, that in this case the gradient-descent method fails to make more progress, due to errors in determining the difference in eqn (7.4).

An obvious issue with the convergence properties of the gradient-descent method is inherent in the update rule, eqn (7.5). Intuitively, we would like to take large steps at locations where the gradient is small (along the valley bottom, for example), and small steps when the gradient is steep (to avoid the problem of overshooting the valley bottom). Unfortunately the update rule eqn (7.5) does the opposite of the intuitive strategy as the increment is proportional to the magnitude of the gradient. The gradient is perpendicular to the contour lines at the location where the gradient is evaluated, see Fig. 7.4. After each iteration the direction of the gradient is orthogonal to the previous step direction and hence the trajectory across the error surface proceeds inefficiently in a zig-zag manner down the valley. Many of the problems associated with the method of steepest descent can be remedied by using knowledge of the curvature of the error surface (the second derivatives) in addition to the gradient.

7.1.4   Second-order expansion: the Newton method

The shape of the error surface in the vicinity of the minimum is approximately parabolic, see Fig. 6.10, therefore a second-order expansion in the parameters about the local minimum is an excellent approximation—this is the Newton method. A Taylor-series expansion of χ^2 about the set of parameters a_s yields[1]

$$\chi^2 (a_s + h) \approx \chi^2 (a_s) + g_s^T h + \frac{1}{2} h^T H_s h. \qquad (7.6)$$

Here the small change in the parameters is the vector h; the **gradient vector** g_s (which has $\mathcal{N}$ components) is

$$g_s = \nabla \chi^2 (a_s) = \left[\frac{\partial \chi^2}{\partial a_1}, \cdots, \frac{\partial \chi^2}{\partial a_{\mathcal{N}}} \right]^T, \qquad (7.7)$$

and the $\mathcal{N} \times \mathcal{N}$ **Hessian matrix** H_s is

$$\mathsf{H}_s = \mathsf{H}(\mathbf{a}_s) = \begin{bmatrix} \frac{\partial^2 \chi^2}{\partial a_1^2} & \cdots & \frac{\partial^2 \chi^2}{\partial a_1 \partial a_{\mathcal{N}}} \\ \vdots & \ddots & \vdots \\ \frac{\partial^2 \chi^2}{\partial a_1 \partial a_{\mathcal{N}}} & \cdots & \frac{\partial^2 \chi^2}{\partial a_{\mathcal{N}}^2} \end{bmatrix}. \tag{7.8}$$

In these equations the superscript T denotes the transpose operation.[2]
At the location of the minimum $\nabla \chi^2 = 0$, thus

$$\nabla \chi^2 (\mathbf{a}_s + \mathbf{h}) = \mathbf{g}_s + \mathsf{H}_s \mathbf{h} = 0, \tag{7.9}$$

which has the solution

$$\mathbf{h} = -\mathsf{H}_s^{-1} \mathbf{g}_s. \tag{7.10}$$

[2]The elements of the transpose of a matrix are related to the elements of the original matrix by the relation $\left(A^{\mathrm{T}}\right)_{jk} = A_{kj}$.

Note that if χ^2 has an exact quadratic dependence on the parameters then the solution of eqn (7.10) is exact, and the problem is solved in one step; else, the process is iterative, and we use the update rule for the next iterate:

$$\mathbf{a}_{s+1} = \mathbf{a}_s + \mathbf{h} = \mathbf{a}_s - \mathsf{H}_s^{-1} \mathbf{g}_s. \tag{7.11}$$

There exist many efficient matrix methods for solving the set of linear equations (7.10). Newton's method has a quadratic convergence (the number of accurate digits of the solution doubles in every iteration). Often the method proceeds with a line minimisation, with a scaled step size (as described in Section 7.1.3). Note that if the Hessian matrix is equal to the unit matrix ($\mathsf{H} = \mathbb{1}$) then this method is equivalent to the method of steepest descent.

7.1.5 Second-order expansion: the Gauss–Newton method

Newton's method involves calculating and inverting the Hessian matrix, which contains second-order derivatives of the χ^2 surface. In the Gauss-Newton method a simpler approximate form of the Hessian is used, as outlined below.
Let us rewrite χ^2 as a sum over the N data points of normalised residuals:

$$\chi^2 = \sum_{i=1}^{N} R_i^2, \tag{7.12}$$

with

$$R_i = \frac{y_i - y(x_i; \mathbf{a})}{\alpha_i}. \tag{7.13}$$

We define the $(N \times \mathcal{N})$ **Jacobian matrix** as follows:

$$\mathsf{J}_s = \mathsf{J}(\mathbf{a}_s) = \begin{bmatrix} \frac{\partial R_1}{\partial a_1} & \cdots & \frac{\partial R_1}{\partial a_{\mathcal{N}}} \\ \vdots & \ddots & \vdots \\ \frac{\partial R_N}{\partial a_1} & \cdots & \frac{\partial R_N}{\partial a_{\mathcal{N}}} \end{bmatrix}. \tag{7.14}$$

Consider the gradient of χ^2 with respect to the parameter a_j:

$$\frac{\partial \chi^2}{\partial a_j} = \frac{\partial}{\partial a_j} \sum_{i=1}^{N} R_i^2 = 2 \sum_{i=1}^{N} \frac{\partial R_i}{\partial a_j} R_i. \qquad (7.15)$$

We see that we can write

$$\nabla \chi^2 = \nabla \sum_{i=1}^{N} R_i^2 = 2 \mathsf{J}^\mathsf{T} \mathsf{R}. \qquad (7.16)$$

Here R is a column vector with N entries for the residuals. The elements of the Hessian matrix are

$$\frac{\partial^2 \chi^2}{\partial a_j \partial a_k} = \frac{\partial^2}{\partial a_j \partial a_k} \sum_i R_i^2 = 2 \sum_i \frac{\partial R_i}{\partial a_j} \frac{\partial R_i}{\partial a_k} + 2 \sum_i R_i \frac{\partial^2 R_i}{\partial a_j \partial a_k}. \qquad (7.17)$$

This equation contains two summations; the first over the product of gradients, the second over second-order derivatives. If the second set of terms is ignored, it is possible to approximate the Hessian in the compact form $\mathsf{H} \approx 2\mathsf{J}^\mathsf{T} \mathsf{J}$. The justification for this approximation comes from looking at the magnitude of the term multiplying the second-order derivative: it is simply the normalised residual R_i. Typically, the normalised residuals should be small, and in the vicinity of $\chi^2_{\min}$ scattered randomly around zero; hence, on being summed, these terms should make a negligible contribution to the Hessian. By throwing away the second order terms the Hessian is much simpler to evaluate, thereby making the method more efficient.[3]

The main advantage of the Gauss–Newton method is that the minimisation requires much fewer steps to converge than the previously discussed methods. The major disadvantage is that the method cannot be relied upon to converge towards the minimum if the initial trial solution is outside the region where the error surface is parabolic. As the second-order expansion methods use the curvature of the error surface the method will not work in regions where the curvature has the wrong sign.

7.1.6   The Marquardt–Levenberg method

A more robust fitting strategy is one in which the best elements of the expansion and gradient approach are combined. Ideally this fitting routine would use the method of steepest descent to progress towards the minimum when the trial solution was far from the optimum values and, as the goodness-of-fit parameter reduces, the minimisation would switch smoothly to an expansion method as the minimum was approached and the surface becomes parabolic.

Building on the insight that the gradient descent and Gauss–Newton methods are complementary, Levenberg suggested the following update rule (Levenberg 1944):

$$\mathsf{a}_{s+1} = \mathsf{a}_s - (\mathsf{H}_s + \lambda \mathbb{1})^{-1} \mathsf{g}_s. \qquad (7.18)$$

The positive constant λ is usually referred to as the damping, or regularisation, constant. Large values of λ are used when the trial solution is far from the

[3] Note that ignoring some of the terms in the definition of the Hessian influences the trajectory across the error surface, but has no bearing on the values of the fit parameters to which the iterative procedure converges.

minimum; in this case eqn (7.18) reverts to the method of steepest descent, eqn (7.5). By contrast, when the trial solution approaches the optimum value, the error surface becomes parabolic, and smaller values of λ are used such that the quadratic convergence properties of the Newton method are deployed.

Marquardt provided the further insight which remedies one of the major drawbacks of the gradient-descent method discussed in Section 7.1.3, namely the zig-zagging along a valley. Marquardt modified the update rule to take into account the magnitude of the curvature, in such a way that large steps are taken in directions with small curvatures, and smaller steps in directions with large curvatures (Marquardt 1963). In the modified update rule the identity matrix of eqn (7.18) is replaced with the diagonal elements of the Hessian matrix:

$$\mathbf{a}_{s+1} = \mathbf{a}_s - \left(\mathbf{H}_s + \lambda \operatorname{diag}[\mathbf{H}_s]\right)^{-1} \mathbf{g}_s. \tag{7.19}$$

Effectively, the elements of the Hessian matrix are modified according to the rule

$$
\begin{aligned}
H_{jj} &\to H_{jj}\left(1 + \lambda\right) \\
H_{jk} &\to H_{jk} \qquad (j \neq k)
\end{aligned}
\tag{7.20}
$$

and for large λ the matrix becomes **diagonally dominant**.

There is a geometrical interpretation of the Marquardt–Levenberg method. At any given point on the error surface away from the minimum the direction of the subsequent steps of the minimisation routine for the gradient and expansion methods are nearly perpendicular. The Marquardt–Levenberg method uses the parameter λ to vary the angle of the trajectory. Far from the minimum the trajectory follows the path of steepest descent and smoothly changes to that of the expansion trajectory as the minimum is approached and the parabolic approximation becomes increasingly valid. A thorough discussion of how to choose initial values for λ and how to update λ during the minimisation procedure can be found in Press *et al.* (1992).

Although more complex, the Marquardt–Levenberg algorithm is clearly superior to any of the previously discussed minimisation routines. It is the minimisation method used in almost all fitting packages and should be used whenever possible.

Table 7.1 provides a summary of the update formulae used by the five algorithms discussed above.

Table 7.1 Formulae used to update the next iterate ($\mathbf{a}_{s+1} = \mathbf{a}_s + \mathbf{h}$) for the five optimisation methods discussed in the text. $\mathbf{g}$ is the gradient vector, $\mathbf{H}$ is the Hessian matrix, and $\mathbf{J}$ is the Jacobian matrix.

Gradient descent	$\mathbf{h} = -\beta\mathbf{g}$
Newton	$\mathbf{H}\mathbf{h} = -\mathbf{g}$
Gauss–Newton	$2\mathbf{J}^T\mathbf{J}\mathbf{h} = -\mathbf{g}$
Levenberg	$(\mathbf{H} + \lambda\mathbf{1})\,\mathbf{h} = -\mathbf{g}$
Marquardt–Levenberg	$\left(\mathbf{H} + \lambda \operatorname{diag}[\mathbf{H}]\right)\mathbf{h} = -\mathbf{g}$

In the final stages of the Marquardt–Levenberg algorithm the matrix used in the update rule approaches the Hessian. After convergence at the desired level has been achieved, the update matrix can be evaluated with $\lambda = 0$; this process yields the Hessian evaluated with the best-fit parameters. The elements of this matrix quantify the curvature of the parabolic error surface in the vicinity of the minimum, and are used to evaluate the uncertainties in the fit parameters.

7.2   The covariance matrix and uncertainties in fit parameters

7.2.1   Extracting uncertainties in fit parameters

The uncertainty of a parameter α_j is defined as the extremum of the $\chi^2_{\text{min}} + 1$ contour along the a_j-axis. As we saw in Chapter 6, when the parameter moves away from its optimal value by an amount equal to $\Delta a = a_j - \bar{a}_j = \alpha_j$, χ^2 evolves to $\chi^2_{\text{min}} + 1$. We now wish to utilise the matrix methods introduced in the previous section to calculate the uncertainties in the $\mathcal{N}$ fit parameters.

It becomes convenient to remove some of the factors of 2 in the earlier discussion by defining the **curvature matrix**, **A**, which is equal to one-half of the Hessian matrix.[4] The off-diagonal terms are related to the degree of correlation of the uncertainties in the parameters, as they describe the curvature of the χ^2 surface along directions which are not collinear with a parameter axis.

The matrix which is the inverse of the curvature matrix is called the **covariance**, or **error matrix**, **C**:

$$[\,\mathsf{C}\,] = [\,\mathsf{A}\,]^{-1}. \tag{7.21}$$

[4]The curvature matrix is also an $\mathcal{N} \times \mathcal{N}$ matrix with components $A_{jk} = \frac{1}{2}\frac{\partial \chi^2}{\partial a_j \partial a_k}$.

The elements of the covariance matrix quantify the statistical errors on the best-fit parameters arising from the statistical fluctuations of the data. If the uncertainties in the measurements are normally distributed then we can extract quantitative information from the error matrix.[5] In Chapter 6 we derived the result for the error in one dimension by analysing the $\Delta\chi^2 = 1$ contour (eqn 6.13); the variance (the uncertainty squared) was seen to be inversely proportional to the curvature of the error surface (eqn 6.12). With $\mathcal{N}$ fit parameters we extend the discussion in Section 6.5.2, such that the contour on the error surface defined by $\chi^2_{\text{min}} + 1$ has tangent planes located at $\pm$ the uncertainties displaced from the optimal values. The matrix equivalent of locating the $\Delta\chi^2 = 1$ contour shown graphically in two dimensions in Fig. 6.8 can be obtained from eqns (7.6) and (7.21); it can be shown (see Press 1992, Section 15.6) that the uncertainty, α_j, in the parameter a_j is

[5]For uncertainties which are not normally distributed one often uses Monte Carlo techniques to ascertain which is the appropriate $\Delta\chi^2$ contour; see Chapter 9.

$$\alpha_j = \sqrt{C_{jj}}. \tag{7.22}$$

This is the most important result for calculating the uncertainties in fit parameters having performed a χ^2 minimisation:

The variance in the jth parameter is given by C_{jj}, the jth diagonal element of the error matrix evaluated with the best-fit paramters.

In the case where the errors are uncorrelated, the off-diagonal terms of the curvature matrix are zero and the diagonal elements of the covariance matrix are simply the inverse of the individual elements of the curvature matrix, i.e. $C_{jj} = (A_{jj})^{-1}$. However, in the more general case where the errors are correlated, the diagonal elements of the error matrix are *not* equal to the inverse of the diagonal elements of the curvature matrix, and the matrix inversion of eqn (7.21) must be performed.

7.2.2   Curvature matrix for a straight-line fit

In the previous section we saw that in the final stages of a Marquardt–Levenberg fitting procedure the curvature matrix is evaluated at the best-fit parameters. When fitting a straight line to a data set the two-dimensional error surface has the same curvature with respect to the slope and intercept for *any* values of these two parameters. Hence, in this special case, the curvature matrix can be calculated quite simply, without even having to perform the least-squares fit.

For the special case of fitting to $y = mx + c$ the four elements of the curvature matrix are given by (see Exercise (7.7) for a derivation)

$$A_{cc} = \sum_i \frac{1}{\alpha_i^2}, \tag{7.23}$$

$$A_{cm} = A_{mc} = \sum_i \frac{x_i}{\alpha_i^2}, \tag{7.24}$$

$$A_{mm} = \sum_i \frac{x_i^2}{\alpha_i^2}. \tag{7.25}$$

The error matrix can thus be found simply by inverting this 2×2 matrix.[6]

[6] Recall that
$$\begin{bmatrix} A & B \\ C & D \end{bmatrix}^{-1} = \frac{1}{AD - BC} \begin{bmatrix} D & -B \\ -C & A \end{bmatrix}.$$

7.2.3   Scaling the uncertainties

Inherent in the discussion so far about extracting error bars on fit parameters is that the model is a good fit to the data. There are circumstances where the value of $\chi^2_{\min}$ obtained is slightly larger than the ideal one. In this case the uncertainties in the parameters are often scaled. This topic is discussed further in Chapter 8.

7.3   Correlations among uncertainties of fit parameters

7.3.1   Correlation coefficients—off-diagonal elements of the covariance matrix

In Chapter 4, when developing a calculus-based approximation for the propagation of errors through multi-variable functions, we discarded the off-diagonal terms, arguing that they were zero for independent variables. If we are to calculate a quantity that is a function of more than one of the fit parameters, our calculation needs to take into account the correlations among the uncertainties extracted from the fit. The magnitude of these correlations is contained within the off-diagonal elements of the covariance matrix, C_{jk}, which are **correlation coefficients** of the fit uncertainties α_j and α_k. Note that the diagonal components C_{jj} are necessarily positive, while the off-diagonal elements C_{jk} can be negative.

We illustrate the concept of covariance of two variables as follows. Consider Z which is a function of two variables A and B, $Z = f(A, B)$. Let there be N pairs of measurements of A and B, (A_i, B_i). For the N measurements of A we can compute the mean, $\overline{A}$, and standard deviation σ_A, and similarly for B. We can also calculate N values of the function $Z_i = f(A_i, B_i)$. Assuming that the errors are small we can use a first-order expansion to find the spread of values of Z_i:

$$Z_i \approx f\left(\overline{A}, \overline{B}\right) + \left(A_i - \overline{A}\right) \frac{\partial Z}{\partial A} + \left(B_i - \overline{B}\right) \frac{\partial Z}{\partial B}. \tag{7.26}$$

It is easy to show that the mean value of Z is given by $\overline{Z} = f\left(\overline{A}, \overline{B}\right)$. We can calculate the sample variance of the N values of Z_i as follows:[7]

$$\sigma_Z^2 = \frac{1}{N-1} \sum_{i=1}^{N} \left(Z_i - \overline{Z}\right)^2, \tag{7.27}$$

then from eqn (7.26) we have:

$$\sigma_Z^2 = \frac{1}{N-1} \sum_{i=1}^{N} \left(\frac{\partial Z}{\partial A}\left(A_i - \overline{A}\right) + \frac{\partial Z}{\partial B}\left(B_i - \overline{B}\right)\right)^2$$

$$\sigma_Z^2 = \left(\frac{\partial Z}{\partial A}\right)^2 \sigma_A^2 + \left(\frac{\partial Z}{\partial B}\right)^2 \sigma_B^2 + 2\frac{\partial Z}{\partial A}\frac{\partial Z}{\partial B}\sigma_{AB}. \tag{7.28}$$

In eqn (7.28) σ_A^2 and σ_B^2 are the variances of A and B, respectively, and we have also introduced the **covariance** σ_{AB}, defined as

$$\sigma_{AB} = \frac{1}{N-1} \sum_{i=1}^{N} \left(A_i - \overline{A}\right)\left(B_i - \overline{B}\right). \tag{7.29}$$

We can extend the discussion of the covariance between two sets of variables to the covariance between fit parameters. Recall that the uncertainty in a parameter we extract from a fit is the standard error, which is the standard deviation of the mean. Hence we can use eqn (7.29) for the propagation of uncertainties in correlated variables, with α to be substituted for σ. Note that the variances and covariance do not necessarily have to have the same units.

It is often easier to use a dimensionless measure of the correlation of two variables, and to this end we introduce the $(N \times N)$ **correlation matrix**. The diagonal elements equal one, and the off-diagonal elements, ρ_{AB}, are called **correlation coefficients**, and are defined as

$$\rho_{AB} = \frac{\sigma_{AB}}{\sigma_A \sigma_B} = \frac{C_{AB}}{\sqrt{C_{AA}C_{BB}}}. \tag{7.30}$$

[7] $N-1$ appears in the denominator, as it did in Chapter 2, on account of the fact that we have one fewer degree of freedom, as the mean is calculated from the data.

Correlation coefficients are dimensionless quantities constrained to the range $-1 \leq \rho_{AB} \leq 1$. If the two variables are uncorrelated then $\rho_{AB} \approx 0$; if ρ_{AB} is close to 1 the variables are strongly positively correlated (a positive value of $A_i - \overline{A}$ is likely to be associated with a positive value of $B_i - \overline{B}$); and if ρ_{AB} is close to -1 the variables are strongly negatively correlated (a positive value of $A_i - \overline{A}$ is likely to be associated with a negative value of $B_i - \overline{B}$).

We illustrate the concepts of covariance and correlations of two variables A and B graphically in Fig. 7.6 which shows a scatter plot of 30 pairs of measurements of (A_i, B_i). In part (a) there is strong positive correlation between the variables ($\rho_{AB} = 0.95$); in (b) there is moderate positive correlation ($\rho_{AB} = 0.60$); in (c) there is very little correlation ($\rho_{AB} = -0.02$); and in (d) there is strong negative correlation ($\rho_{AB} = -0.81$).

The off-diagonal element C_{jk} of the covariance matrix contains information about the correlation between the uncertainties in the parameters a_j and a_k. This information can be incorporated into error-propagation calculations, as we show in the next section.

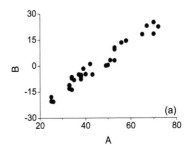

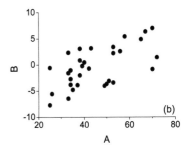

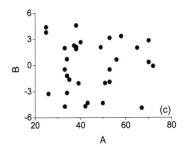

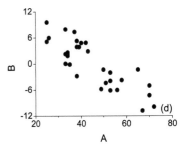

Fig. 7.6 An illustration of the linear correlation between two variables A and B. In (a) there is strong positive correlation between the variables ($\rho_{AB} = 0.95$); in (b) there is moderate positive correlation ($\rho_{AB} = 0.60$); in (c) there is very little correlation ($\rho_{AB} = -0.02$); and in (d) there is strong negative correlation ($\rho_{AB} = -0.81$).

7.4 Covariance in error propagation

In this section we provide a look-up table for some common functions of two correlated variables, and we illustrate the concepts introduced in the last section by an example, including the very important case of the calibration curve.

Table 7.2 Some rules for the propagation of errors with two correlated variables.

Function, $Z(A, B)$	Expression used to calculate α_Z
$Z = A \pm B$	$\alpha_Z^2 = \alpha_A^2 + \alpha_B^2 \pm 2\alpha_{AB}$
$Z = A \times B$	$\left(\frac{\alpha_Z}{Z}\right)^2 = \left(\frac{\alpha_A}{A}\right)^2 + \left(\frac{\alpha_B}{B}\right)^2 + 2\left(\frac{\alpha_{AB}}{AB}\right)$
$Z = \frac{A}{B}$	$\left(\frac{\alpha_Z}{Z}\right)^2 = \left(\frac{\alpha_A}{A}\right)^2 + \left(\frac{\alpha_B}{B}\right)^2 - 2\left(\frac{\alpha_{AB}}{AB}\right)$

7.4.1 Worked example 1—a straight-line fit

In Fig. 6.9 in Chapter 6 we saw the χ^2 contours for the two parameters, slope and intercept. As the iso-χ^2 ellipses are tilted we expect to find a correlation between the uncertainties of the fit parameters. Using eqns (7.24)–(7.25) from Section 7.2.2, we calculate the curvature matrix A to be:

$$\mathsf{A} = \begin{bmatrix} 0.362 \ (\text{mV})^{-2} & 20.6 \ \text{Hz}(\text{mV})^{-2} \\ 20.6 \ \text{Hz}(\text{mV})^{-2} & 1538 \ (\text{Hz})^2(\text{mV})^{-2} \end{bmatrix}.$$

By inverting the curvature matrix we obtain the error matrix:

$$C = \begin{bmatrix} 11.5 \text{ (mV)}^2 & -0.153 \text{ (mV)}^2\text{Hz}^{-1} \\ -0.153 \text{ (mV)}^2\text{Hz}^{-1} & 0.0027 \text{ (mV)}^2\text{(Hz)}^{-2} \end{bmatrix}.$$

The square roots of the diagonal elements (to one significant figure) are 3 mV and 0.05 mV/Hz, which are the uncertainties in the intercept and slope, respectively, reported in Table 6.1. The correlation matrix is

$$\begin{bmatrix} 1.00 & -0.871 \\ -0.871 & 1.00 \end{bmatrix}.$$

The negative off-diagonal element reflects the fact that the ellipse is tilted in such a way that, for a given value of χ^2, an increase in gradient is accompanied by a decrease in intercept. Let us now use these fit parameters to predict the expected value, V, of the voltage, and its error, at a frequency, f, of 75 Hz. The quantities are related by the expression $V = mf + c$, thus $V = 2.03 \times 75 - 1 = 151$ mV. If we were to ignore the correlation of the uncertainties the error in V would be calculated using the first entry in Table 4.2 as

$$\alpha_V^2 = f^2\alpha_m^2 + \alpha_c^2 = f^2 C_{22} + C_{11}, \tag{7.31}$$

where we have made use of the fact that the variances in the parameters are equal to the diagonal elements of the error matrix. Inserting numerical values, we obtain $\alpha_V = \sqrt{75^2 \times 0.0027 + 11.5} = 5$ mV.

Incorporating the correlation into our calculation using the first entry of Table 7.2 yields a different result:

$$\alpha_V^2 = f^2 C_{22} + C_{11} + 2f C_{12}. \tag{7.32}$$

Note the presence of the covariance term in eqn (7.32). Inserting numerical values, $\alpha_V = \sqrt{75^2 \times 0.0027 + 11.5 + 2 \times 75 \times -0.153} = 2$ mV. With respect to the result obtained without the correlation term, this is (a) different, and (b) reduced—a consequence of the correlation being negative.[8]

[8]Note that the value of the predicted voltage (151 mV in this case) does not change if the correlations in the uncertainties are included.

7.4.2   Worked example 2—a four-parameter fit

We saw qualitatively for the data plotted in Fig. 6.11 that the values of the background and peak centre extracted from the fit were largely uncorrelated, whereas the values of the background and peak width were strongly correlated. We can analyse the degree of correlation for the fit more quantitatively by analysing the terms in the correlation matrix. In this example, the correlation matrix is

$$\begin{bmatrix} 1.00 & -0.64 & 0.94 & -0.25 \\ -0.64 & 1.00 & -0.53 & 0.0893 \\ 0.94 & -0.53 & 1.00 & -0.297 \\ -0.25 & 0.0893 & -0.297 & 1.00 \end{bmatrix}.$$

Here, the first parameter is the amplitude, the second the background, the third the peak width, and the fourth the centre. As expected the correlation of the background and peak centre is small, $\rho_{24} = 0.089$; and the correlation of the background and peak width is strong and negative, $\rho_{23} = -0.53$. Note also that as $\rho_{13} = 0.94$ there is a strong positive correlation between the amplitude and peak width—one can trade off an increase in the peak height and width by reducing the background.

Chapter summary

- Computer codes find the minimum of χ^2 by the use of iterative algorithms for the fit parameters.
- In the vicinity of a minimum of χ^2, the gradient of χ^2 with respect to the fit parameters is zero.
- Far from the minimum a method based on knowledge of the gradient of the error surface with respect to the fit parameters (the method of steepest descent) can be used to head in the direction of the minimum.
- In the vicinity of the minimum, the error surface is a quadratic function of the changes of parameters from their optimal values, and second-order (Newton) methods based on quadratic expansion are used.
- The behaviour of χ^2 in the vicinity of a minimum is governed by the second-order derivatives contained in the Hessian matrix, H.
- The Marquardt–Levenberg method combines the best features of gradient descent and expansion, and is ubiquitous in minimisation algorithms.
- The curvature matrix, A, is equal to one-half of the Hessian matrix.
- The error matrix, C, is the inverse of the curvature matrix.
- The variance in the jth parameter is given by C_{jj}, the jth diagonal element of the error matrix.
- The off-diagonal elements of the error matrix contain information about the correlation of the uncertainties of the fit parameters.
- The linear correlation coefficient ρ_{ij} quantifies the correlation between the ith and j^{th} parameters.

Exercises

(7.1) *Convergence properties of Newton's method*
We investigate the convergence properties of the Newton–Raphson method to find zeroes of a function presented in Section 7.1.1 by example. Let us find an approximation to $\sqrt{26}$ to 10 decimal places. This process is equivalent to finding the zero crossing of the function $f(x) = x^2 - 26$. We use the equation

$$x_{s+1} = x_s - \frac{f(x_s)}{f'(x_s)} = x_s - \frac{(x_s)^2 - 26}{2x_s},$$

where the derivative of the function, $f'(x) = 2x$, has been substituted. (a) As $\sqrt{25} = 5$, we start with the guess of $x_1 = 5$. Show that the next iterations are: $x_2 = 5.1, x_3 = 5.099\,019\,607\,843, x_4 = 5.099\,019\,513\,593$, and $x_5 = 5.099\,019\,513\,593$. Note how convergence to better than 10 decimal places is achieved with only five iterations. (b) What solution do you obtain with an initial guess of $x_1 = -5$? Explain your answer. (c) What solution do you obtain with an initial guess of $x_1 = 0$? Explain your answer.

(7.2) *Newton's method for finding maxima and minima*

Modify the results of Section 7.1.1 to derive the result for the approximate location, x_{s+1}, of the maximum or minimum of a function, given the value of the first, $f'(x_s)$, and second derivatives, $f''(x_s)$, of the function at the current location, x_s:

$$x_{s+1} = x_s - \frac{f'(x_s)}{f''(x_s)}.$$

(7.3) *Finding the minimum of a parabola from three points*

Consider the variations in χ^2 with respect to the parameter a_k. Let the step size be Δ. The grid-search method proceeds with χ^2 reducing as more step sizes are made, until the first occurrence of an increase in χ^2. The last two values of a_k must bracket the value at which χ^2 is minimised. Using the last three values it is possible to calculate the value where the minimum occurs. Let the three values of the parameter be $\tilde{a}_k$, $\tilde{a}_k + \Delta$ and $\tilde{a}_k + 2\Delta$, and the values of χ^2 for those parameters be χ_1^2, χ_2^2 and χ_3^2, respectively. (a) Show that the minimum χ^2 is achieved at a value of

$$a_k = \tilde{a}_k + \frac{\Delta}{2}\left(\frac{3\chi_1^2 - 4\chi_2^2 + \chi_3^2}{\chi_1^2 - 2\chi_2^2 + \chi_3^2}\right),$$

assuming a parabolic dependence of χ^2 on a_k near the minimum. (b) Recalling that the uncertainty, α_k, in the parameter a_k is ascertained by finding the increase from the minimum which will increase χ^2 by 1, show that

$$\alpha_k = \Delta\sqrt{\frac{2}{\chi_1^2 - 2\chi_2^2 + \chi_3^2}}.$$

(7.4) *Error propagation for correlated errors*

Assume in this question that the uncertainties in A and B are correlated. Verify the results in Table 7.2. (i) If $Z = A \pm B$, show that $\alpha_Z^2 = \alpha_A^2 + \alpha_B^2 \pm 2\alpha_{AB}$.
(ii) If $Z = A \times B$, show that
$\left(\frac{\alpha_Z}{Z}\right)^2 = \left(\frac{\alpha_A}{A}\right)^2 + \left(\frac{\alpha_B}{B}\right)^2 + 2\left(\frac{\alpha_{AB}}{AB}\right)$.
(iii) If $Z = \frac{A}{B}$, show that
$\left(\frac{\alpha_Z}{Z}\right)^2 = \left(\frac{\alpha_A}{A}\right)^2 + \left(\frac{\alpha_B}{B}\right)^2 - 2\left(\frac{\alpha_{AB}}{AB}\right)$.

(7.5) *Geometry of error calculation in two dimensions*

In two dimensions the covariance matrix is written

$$C = \begin{bmatrix} \alpha_1^2 & \rho\alpha_1\alpha_2 \\ \rho\alpha_1\alpha_2 & \alpha_2^2 \end{bmatrix}.$$

Show that the inverse of the covariance matrix can be cast in the following form:

$$C^{-1} = \frac{1}{1-\rho^2}\begin{bmatrix} \frac{1}{\alpha_1^2} & -\frac{\rho}{\alpha_1\alpha_2} \\ -\frac{\rho}{\alpha_1\alpha_2} & \frac{1}{\alpha_2^2} \end{bmatrix}.$$

Show that the two eigenvalues $\lambda_\pm$ of the covariance matrix are

$$\lambda_\pm = \frac{1}{2}\left(\alpha_1^2 + \alpha_2^2 \pm \sqrt{\left(\alpha_1^2 + \alpha_2^2\right)^2 + 4\rho^2\alpha_1^2\alpha_2^2}\right).$$

The two orthonormal eigenvectors can be written as $\begin{pmatrix}\cos\phi \\ \sin\phi\end{pmatrix}$ and $\begin{pmatrix}-\sin\phi \\ \cos\phi\end{pmatrix}$, where ϕ is the angle of the major axis of the $\chi^2 + 1$ contour. Show that the angle ϕ is given by

$$\tan 2\phi = \frac{2\rho\alpha_1\alpha_2}{\alpha_1^2 - \alpha_2^2}.$$

(7.6) *Terms in the correlation matrix*

Show that the diagonal terms of the correlation matrix are equal to unity.

(7.7) *Curvature matrix for a straight-line fit*

Recalling that $\chi^2 = \sum_i \frac{(y_i - y(x_i))^2}{\alpha_i^2}$, and that after an experiment has been performed the values of x_i, y_i and α_i are fixed, show that

(i) $\frac{1}{2}\frac{\partial^2\chi^2}{\partial c^2} = \sum_i \frac{1}{\alpha_i^2}$,

(ii) $\frac{1}{2}\frac{\partial^2\chi^2}{\partial m\partial c} = \sum_i \frac{x_i}{\alpha_i^2}$,

(iii) $\frac{1}{2}\frac{\partial^2\chi^2}{\partial m^2} = \sum_i \frac{x_i^2}{\alpha_i^2}$.

Hence verify the results given for the elements of the curvature matrix in Section 7.3.

(7.8) *Using a calibration curve*

A frequently encountered case where the correlation of the uncertainties must be taken into account is that of a **calibration curve**. Consider the following set of measurements from an optical-activity experiment, where the angle of rotation of a plane-polarized light beam, θ, is measured as a function of the independent variable, the concentration, C, of a sucrose solution.

C (g cm^{-3})	0.025	0.05	0.075	0.100
θ (degrees)	10.7	21.6	32.4	43.1

C (g cm^{-3})	0.125	0.150	0.175
θ (degrees)	53.9	64.9	75.4

The errors in the angle measurement are all $0.1°$, the errors in the concentration are negligible. A straight-line fit to the data yields a gradient of $431.7°\ \mathrm{g}^{-1}\ \mathrm{cm}^3$, and intercept $-0.03°$. Show that the curvature matrix, A, is given by

$$A = \begin{bmatrix} 700\ \left((°)^{-2}\right) & 70\ \left((°)^{-2}\mathrm{g\ cm}^{-3}\right) \\ 70\ \left((°)^{-2}\mathrm{g\ cm}^{-3}\right) & 8.75\ \left((\mathrm{g/°\ cm}^3)^2\right) \end{bmatrix}.$$

and that the error matrix is

$$C = \begin{bmatrix} 0.00714\ \left((°)^2\right) & -0.0571\ \left((°)^2\mathrm{g}^{-1}\mathrm{cm}^3\right) \\ -0.0571\ \left((°)^2\mathrm{g}^{-1}\mathrm{cm}^3\right) & 0.571\ \left((°)^2\mathrm{g}^{-2}\ \mathrm{cm}^6\right) \end{bmatrix}.$$

The entry for the intercept is in the top left-hand corner, that for the gradient in the bottom right-hand corner. Calculate the associated correlation matrix. Use the entries of the error matrix to answer the following questions: (i) what are the uncertainties in the best-fit intercept and gradient? (ii) what optical rotation is expected for a known concentration of $C = 0.080\ \mathrm{g\ cm}^{-3}$, and what is the uncertainty? and (iii) what is the concentration given a measured rotation of $\theta = 70.3°$ and what is the uncertainty?

Hypothesis testing – how good are our models?

8

In Chapter 5 we introduced the goodness-of-fit parameter, χ^2. We have seen in the preceding chapters how minimisation of this statistic yields both the best fit of a model to a data set and the uncertainties in the associated parameters. Implicit in the discussion so far has been the assumption that the model was a correct description of the data. We have shown how various methods can be used to answer the question 'what is the best fit to my data?'. However, there is often a far more interesting question: 'are my data consistent with the proposed model?'. For instance, 'is the best fit to the data a straight line?'.

The subject of this chapter is hypothesis testing in the context of error analysis. We will show how it is possible to use statistical tests to give a probability that a particular hypothesis is valid. In earlier chapters we presented techniques for reporting the best estimate of a parameter with its associated uncertainty. We have shown that this uncertainty is a probabilistic statement of a confidence limit, based on an appropriate theoretical model. In this chapter we extend this idea and apply statistical tests to hypotheses or concepts.

8.1 Hypothesis testing

There exist several statistical tests that calculate the probability that the data may be described by a given model. This is undertaken by defining the **null hypothesis**, H_0, which is often the assumption that an obtained sample distribution can be described by a particular parent distribution. When we extend these arguments to testing the quality of a particular fit we test the null hypothesis that our data, y_i (the sample distribution), is well modelled by a particular function, $y(x_i)$ (the parent distribution). Statistical techniques allow the null hypothesis to be tested quantitatively to determine whether the null hypothesis should be rejected and an alternative hypothesis pursued. The default in fitting a model to a data set is that at a particular confidence level or probability there is no evidence that the null hypothesis should be rejected. We note that the null hypothesis is never accepted as there always remains a finite probability that an alternative hypothesis represented by a different parent distribution would be a better description of the data and hence yield a better fit.

A statistic that is commonly used to test the significance of the null hypothesis in the physical sciences is the χ^2 statistic. The sample distribution is tested against a particular parent distribution taking into account the experimental uncertainties. We have already seen in Chapters 6 and 7 that minimising the χ^2 statistic allows the best-fit model parameters and their uncertainties to be estimated. The resulting minimum value of the χ^2 statistic, χ^2_{min}, is a numerical measure of the discrepancy between the proposed model and the data. As the χ^2 statistic is summed over all the data points it is a quantitative measure of how well the entire data set can be modelled with the proposed parent distribution and, as such, can be used as a test of the null hypothesis.

For any given data set we can calculate the probability of obtaining a value of χ^2 equal to χ^2_{min} or higher, given the proposed model.[1] When this probability is sufficiently low (5% and 1% are frequently used) we would reject the hypothesis at the appropriate percentage level.

[1] In this case we compare the null hypothesis that the data are well modelled by the function against the alternative that the statistical variation in the data set is random.

8.2   Degrees of freedom

We have seen in preceding chapters that it is possible to give estimates for various statistical parameters based on different numbers of data points in a set of measurements. The number of unconstrained variables is known as the number of **degrees of freedom**, and represents the number of independent pieces of information that are used to evaluate an estimate of a parameter. In general, the degrees of freedom is equal to the number of independent data points used in the estimate minus the number of parameters used in the estimation that have already been determined from the data set. When fitting N independent data points with a function with $\mathcal{N}$ parameters the number of degrees of freedom, ν, is:

$$\nu = N - \mathcal{N}. \tag{8.1}$$

The more data points that are unconstrained, the more robust a statistical estimate of parameters such as the mean, variance and χ^2 become. We can also consider the degrees of freedom as the number of measurements exceeding the bare minimum necessary to measure a certain quantity. This is a topic which can cause much confusion, largely as there are instances where it is not clear whether one should have many, or few, degrees of freedom. When a complex situation is being analysed, one often makes simplifying approximations, such as restricting the motion to one dimension. The motivation is often 'to reduce the number of degrees of freedom', with the implication being that the smaller the number of degrees of freedom the better. Whereas this can represent the case for simplifying a physical situation before a mathematical description is adopted, in the case of data analysis the opposite is the case—the more degrees of freedom the better.

Consider the case of a measurement of a certain quantity comprising a single data point. It is obviously possible to fit to the hypothesis that the quantity is constant, but no real information is gained from this hypothesis. For two data points, one can ask the more meaningful question 'are the data consistent with a constant?', but it is pointless to ask the question 'are the data consistent with a straight line?' as it is *always* possible to find a straight line that goes through two points. There is very little value in modelling a sample distribution with N data points with a parent distribution with greater than N parameters. In data analysis it is more convincing to use as few constraints as possible when fitting a data set, thus maximising the number of degrees of freedom. We will discuss in Section 8.7 the concept of what constitutes enough constraints in a hypothesis test.

Each parameter of the parent distribution estimated from the sample distribution reduces the number of unconstrained data points by 1. In Section 2.2 we saw that the number of data points, N, appears in the denominator of the definition of the mean (eqn 2.1). However $N - 1$ appears in the denominator of the definition of the standard deviation of a sample (Section 2.3.2, eqn 2.2), as the mean had to be estimated from the same N data points, leaving only $N - 1$ unconstrained values. Moreover, eqn (5.6) used to evaluate the common uncertainty in linear regression in Section 5.2.1 has the factor $N - 2$ in the denominator as both the mean and intercept were calculated from the N data points.

8.2.1 Data reduction and the number of degrees of freedom

We have emphasised throughout this book that reducing a (potentially very large) number of measurements to a handful of parameters and their errors is the *modus operandi* of data analysis. Here we reinforce the concept that having a large number of degrees of freedom is the ideal case. Consider the following quiz question: 'what number comes next in the sequence 1, 2, 4, 6 and 10?'. The 'official answer' is 12. The reasoning is that the series is the sequence of prime numbers less 1. Consider the following function

$$f(x) = 5 - 8.5833x + 5.875x^2 - 1.4167x^3 + 0.125x^4. \qquad (8.2)$$

This function has been designed to have the property $f(1) = 1$, $f(2) = 2$, $f(3) = 4$, etc. For this function $f(6) = 21$, therefore we could argue with the quiz setter about the validity of their answer. By choosing a fourth-order polynomial with five coefficients we can describe the five terms of the sequence exactly—there are no degrees of freedom. Obviously, it is impossible to do this in general with a lower-order polynomial. However, we now need to keep all five coefficients to describe the 5 terms in the sequence—this is very inefficient, and the opposite of data reduction.

We can also use this example to highlight some of the issues with interpolation and extrapolation. There is only one fourth-order polynomial, $f(x)$ in eqn (8.2), which fits the sequence exactly. However there are an infinite number of fifth-order polynomials which fit the five terms of the sequence. Two of them are

$$g(x) = 14 - 29.133\,x + 22.75\,x^2 - 7.7917\,x^3 + 1.25\,x^4 - 0.075\,x^5,$$

$$h(x) = 26 - 56.5333\,x + 45.25\,x^2 - 16.2917\,x^3 + 2.75\,x^4 - 0.175\,x^5.$$

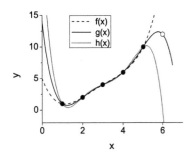

Fig. 8.1 The first six numbers in the sequence 'prime number less one' are shown as solid circles. A fourth-order fit to the first five data points is shown, $f(x)$, in addition to two fifth-order fits, $g(x)$ and $h(x)$. Note also that although the curves show excellent agreement over the range of x values for which the data are defined, they deviate substantially outside this range, highlighting the difficulties inherent in extrapolation.

[2] The gamma function is related to the factorial function. For positive integers n the gamma function is equivalent to $\Gamma(n+1) = n!$, and for positive half-integers it is defined as $\Gamma(n+1/2) = \sqrt{\pi}\,\frac{(2n)!}{n!\,2^{2n}}$.

[3] The **median**, m, of a continuous distribution function, $P_{DF}(x)$, is the value of x for which the probability of finding a value of $x > m$ is equal to the probability of finding $x < m$, i.e.

$$\int_{-\infty}^{m} P_{DF}(x)\,dx = \int_{m}^{\infty} P_{DF}(x)\,dx.$$

The **mode** of a probability function is the value of x for which $P_{DF}(x)$ is a maximum.

These functions have the desired property of being equal to the original sequence for integer arguments, and also $g(6) = 12$, and $h(6) = 0$. Therefore if we extrapolate the function $g(x)$ we see it agrees with the 'prediction' of the quiz question; whereas if we extrapolate $h(x)$ we get a radically different answer. As seen in Fig. 8.1 both functions are smooth, and pass through the 'data', yet they predict very different behaviour outside the range over which we have data. By contrast, the difference when interpolating is smaller: for example $g(3.5) = 4.84$, and $h(3.5) = 4.70$.

8.3   The χ^2 probability distribution function

As the χ^2 statistic is a random variable it also has a normalised probability distribution function, given by (Bevington and Robinson 2003, Chapter 11 and Squires 2001, Appendix E):

$$X\left(\chi^2; \nu\right) = \frac{\left(\chi^2\right)^{\left(\frac{\nu}{2}-1\right)} \exp\left[-\chi^2/2\right]}{2^{\nu/2}\,\Gamma\left(\nu/2\right)}, \tag{8.3}$$

where $\Gamma(x)$ is the gamma function[2] and ν is the number of degrees of freedom. It can be shown using the identities introduced in Chapter 3 for probability distribution functions that $X\left(\chi^2; \nu\right)$ has an *expectation value*, or mean, of ν with a standard deviation of $\sigma_{\chi^2} = \sqrt{2\nu}$. As the χ^2 probability distribution function is asymmetric it is worth noting that the median and mode do not have the same value as the mean: $X\left(\chi^2; \nu\right)$ has a median of approximately $\nu - 2/3$ and a mode equal to $\nu - 2$ for $\nu > 2$, as seen in Fig. 8.2(a).[3]

As with other probability distribution functions, the probability of obtaining a value of χ^2 between χ^2_{min} and ∞ is given by the cumulative probability function, $P\left(\chi^2_{min}; \nu\right)$:

$$P\left(\chi^2_{min} \leq \chi^2 \leq \infty; \nu\right) = \int_{\chi^2_{min}}^{\infty} X\left(\chi^2; \nu\right) d\chi^2. \tag{8.4}$$

Equation (8.4) gives the probability that were the sample distribution drawn from the hypothesised parent distribution, one would obtain a value of χ^2 equal to, or greater than, χ^2_{min}.

Fortunately the cumulative distribution function $P\left(\chi^2_{min}; \nu\right)$ is accessible in most spreadsheet and statistical packages and values for specific combinations of χ^2 and ν are tabulated in many statistical books. The two functions $X\left(\chi^2; \nu\right)$ and $P\left(\chi^2_{min}; \nu\right)$ are plotted in Fig. 8.2.

The forms of both functions are mathematically non-trivial due to the presence of the gamma function and can be difficult to evaluate as $\chi^2 \to 0$. The asymmetry of the χ^2 distribution for low ν is clear in Fig. 8.2(a). This asymmetry is described by the degree of **skewness** in the function, which for the χ^2 distribution reduces slowly as the number of degrees of freedom increases. The corresponding cumulative probability distributions for the χ^2 distributions in Fig. 8.2(a) are shown in Fig. 8.2(b).

8.3.1 χ^2 for one degree of freedom

For one degree of freedom, eqn (8.3) takes the simpler form (Bevington and Robinson 2003, p. 197):

$$X\left(\chi^2;\ 1\right) = \frac{\exp\left[-\chi^2/2\right]}{\sqrt{\left(2\pi\chi^2\right)}}. \tag{8.5}$$

In Chapter 6 we saw that to extract the uncertainty in a parameter one needs to reduce an $\mathcal{N}$-dimensional error surface to a single one-dimensional slice that maps the shape of the error surface with respect to one of the parameters in the vicinity of $\chi^2_{\min}$ with all remaining parameters re-optimised to minimise χ^2. The variation in χ^2 around the minimum, $\Delta\chi^2$, must obey the χ^2 distribution for one degree of freedom (eqn 8.5). The cumulative probability distribution for one degree of freedom can be used to find the probability of obtaining a value of $\Delta\chi^2$ of, say, 1 or higher. In Fig. 8.3 we show the percentage probability of obtaining a value of χ^2 of less than or equal to $(\chi^2_{\min} + \Delta\chi^2)$ as a function of $\Delta\chi^2$. The values in Table 6.2 are obtained from the curve shown in Fig. 8.3. The probability of obtaining a value of $\Delta\chi^2$ up to one is 68%. The other confidence limits for $\Delta\chi^2$ in Table 6.2 are shown as solid points in Fig. 8.3.

8.4 Using χ^2 as a hypothesis test

We can use eqns (8.3) and (8.4) to perform our hypothesis test. We expect that if the proposed model is in good agreement with the data that $\chi^2_{\min}$ will be close to the *mean* of the χ^2 distribution and so $\chi^2_{\min} \approx \nu$. For many degrees of freedom the χ^2 distribution becomes more symmetric and the median, mode and mean become similar and we would expect for a good match between sample and parent distributions a corresponding probability $P\left(\chi^2_{\min} \approx \nu;\ \nu\right) \approx 0.5$. We learn about the quality of the agreement between sample and parent by analysing $P\left(\chi^2_{\min};\ \nu\right)$, the probability of obtaining the observed value, or higher, of $\chi^2_{\min}$ for a given number of degrees of freedom.

If the value of $\chi^2_{\min}$ is significantly greater than ν the probability, $P\left(\chi^2_{\min} > \nu;\ \nu\right)$, is small, as seen in Fig. 8.2(b). Under such circumstances, there are discrepancies between the model and the data which are unlikely to be explained by random statistical fluctuations in the sample distribution. This could arise from either (i) an incorrect null hypothesis, or (ii) an incorrect evaluation or assumption about the uncertainties. Conversely, if the value of $\chi^2_{\min}$ is less than ν the cumulative probability function, $P\left(\chi^2_{\min};\ \nu\right)$, becomes greater than 0.5 and tends towards unity. This is *not* an indication of an improved fit, but rather suggests that the standard errors used in the determination of the χ^2 statistic have been overestimated, resulting in an unrealistically small value of χ^2.

We now turn to the question of when to reject the hypothesis on the grounds of the value of $\chi^2_{\min}$ being too large to attribute to statistical fluctuations in

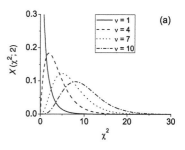

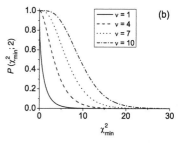

Fig. 8.2 (a) The normalised χ^2 probability distribution function for $\nu = 1,\ 4,\ 7$ and 10 with (b) the corresponding cumulative probability functions.

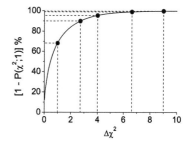

Fig. 8.3 The percentage probability of obtaining a value of χ^2 of $\leq \chi^2_{\min} + \Delta\chi^2$ as a function of $\Delta\chi^2$ for one degree of freedom. The confidence limits quoted in Table 6.2 are shown as solid points.

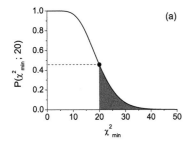

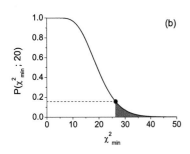

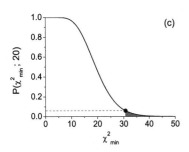

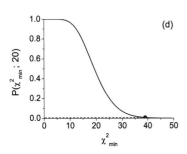

Fig. 8.4 Cumulative probability distribution functions for 20 degrees of freedom. Part (a) shows the case of $\chi^2_{\min}$ equal to the expected value (v), the shaded area represents the probability of obtaining a value of χ^2 equal to v or higher and is 46% of the area under the curve. The cases for $\chi^2_{\min} = v + \sigma$, $v + 2\sigma$ and $v + 3\sigma$ are shown in (b), (c) and (d) respectively, with corresponding probabilities of 16%, 4% and 0.7%.

the data set. Obviously, there is not one critical value of $P\left(\chi^2_{\min}; v\right)$ which determines whether the hypothesis is rejected; in this treatment we will investigate the evolution of $P\left(\chi^2_{\min}; v\right)$ as a function of the results being two or three standard deviations from the mean. In statistics texts (see, e.g. Rees 1987) the 5% level of significance is often chosen as the critical value; i.e. if $P\left(\chi^2; v\right) \le 0.05$ the hypothesis is said 'to be rejected at the 5% level'. As with the sampling of any probability distribution function the most probable value returned will be the expectation value or mean. However, from the discussion in Chapters 2 and 3 we would not be surprised to find results that are within two standard deviations of the mean. For a test of a particular null hypothesis, we would therefore expect to obtain a value of $\chi^2_{\min} \approx v$ but would not be surprised to find $\chi^2_{\min}$ within two standard deviations of the mean, i.e. in the range $v - 2\sqrt{2v} \le \chi^2_{\min} \le v + 2\sqrt{2v}$. The probability of obtaining a value of $\chi^2_{\min}$ of $v + 2\sqrt{2v}$ or higher is approximately 4×10^{-2}. Due to the asymmetry of the χ^2 distribution this probability has a very weak dependence on v; for five degrees of freedom, the probability derived from eqn (8.4) is $P(11.3; 5) = 0.045$; for $v = 20$, $P(32.6; 20) = 0.037$ and for $v = 100$, $P(128; 100) = 0.029$. The probability of finding a value of $\chi^2_{\min}$ greater than three standard deviations from the mean, i.e. $\chi^2_{\min} > v + 3\sqrt{2v}$, is an order of magnitude smaller and is approximately 5×10^{-3}.

We illustrate these ideas in Fig. 8.4 where the cumulative probability distribution for χ^2 with $v = 20$ is shown. In Fig. 8.4(a) the distribution is shown with the mean marked. The probability of obtaining a value of $\chi^2_{\min}$ of 20 or higher with $v = 20$ is $P(20; 20) = 0.46$ which is slightly less than 1/2. In subsequent plots the probabilities of obtaining a value of $\chi^2_{\min}$ of $v + \sigma$, $v + 2\sigma$ and $v + 3\sigma$ or higher are shown. The probabilities for these points are $P(26.3; 20) = 0.16$, $P(32.6; 20) = 0.04$ and $P(39.0; 20) = 0.007$ respectively.

Although the probability of finding a value of $\chi^2_{\min}$ greater than $v + 3\sigma$ is rather low, $P(v + 3\sigma; v) \approx 10^{-3}$, genuinely incorrect models will give significantly lower probabilities, say $P\left(\chi^2_{\min}; v\right) \approx 10^{-18}$. It is for the experimenter to decide at which threshold the null hypothesis is rejected. Although fits can be accepted at the $P\left(\chi^2_{\min}; v\right) \approx 10^{-3}$ level, it is not advisable to do this as a matter of course. It would be a better strategy to find the origin of the discrepancy between the model and data.

- For a reasonable fit, the value of $P\left(\chi^2_{\min}; v\right) \approx 0.5$.
- If $P\left(\chi^2_{\min}; v\right) \to 1$ check your calculations for the uncertainties in the measurements, α_i.
- The null hypothesis is generally **not rejected** if the value of $\chi^2_{\min}$ is within $\pm 2\sigma$ of the mean, v, i.e. in the range $v - 2\sqrt{2v} \le \chi^2_{\min} \le v + 2\sqrt{2v}$.
- The null hypothesis is **questioned** if $P\left(\chi^2_{\min}; v\right) \approx 10^{-3}$ or $P\left(\chi^2_{\min}; v\right) > 0.5$.
- The null hypothesis is **rejected** if $P\left(\chi^2_{\min}; v\right) < 10^{-4}$.

8.4.1 The reduced χ^2 statistic

In order to ascertain whether a particular null hypothesis should be rejected at a particular confidence level, the probabilities of obtaining the observed value of χ^2_{min} or higher given the number of degrees of freedom must be calculated using eqn (8.4). However, we can obtain a quick indication as to whether the null hypothesis should be rejected by considering the so-called **reduced chi-squared statistic**, χ^2_ν, which is simply the value of χ^2_{min} divided by the number of degrees of freedom:

$$\chi^2_\nu = \frac{\chi^2_{min}}{\nu}. \tag{8.6}$$

A good match between the sample and parent distribution occurs when $\chi^2_\nu \approx 1$. We can understand why the null hypothesis is not rejected if $\chi^2_\nu \approx 1$ by considering that for good agreement between the sample and parent distribution each point will differ from its expected value, i.e. the mean, by typically the standard deviation of the mean (standard error). Thus each term in the summation of the χ^2 statistic should be of order one, with the result that $\chi^2_{min} \approx N$. Typically, as the number of degrees of freedom is similar to the number of data points, χ^2_ν is thus expected to be unity. If the value of χ^2_ν is much larger than one, it is likely that the null hypothesis should be rejected. A very small value of χ^2_ν is also unlikely—either the error bars have been overestimated, which reduces the value of χ^2, or the observed and expected values are unrealistically close. Again it is for the experimenter to decide at what confidence limit the null hypothesis should be rejected in terms of the reduced chi-squared statistic, χ^2_ν. In Section 8.4 we discussed how one would not be surprised if the observed value of χ^2_{min} was within 2σ of the mean for a good fit, and that the null hypothesis should only be questioned if the value of χ^2_{min} was larger than, say, 3σ from the mean. As the standard deviation of the χ^2 distribution depends on ν, the confidence limits for χ^2_ν also depend on ν. In Fig. 8.5 the values of χ^2_ν calculated at $\nu + \sigma$, $\nu + 2\sigma$, and $\nu + 3\sigma$ are plotted as a function of the number of degrees of freedom. Recall that for a Gaussian distribution these intervals correspond to the 68%, 95% and 99.7% confidence limits. The 3σ confidence limit of χ^2_ν is tabulated for several values of ν in Table 8.1.

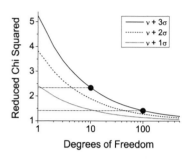

Fig. 8.5 χ^2_ν calculated at $\nu + \sigma$ (dotted), $\nu + 2\sigma$ (dashed), and $\nu + 3\sigma$ (solid) and plotted as a function of the number of degrees of freedom. For 100 degrees of freedom the limit for not rejecting the null hypothesis is for a value of χ^2_ν of 1.4, whereas it is 2.3 for 10 degrees of freedom.

Table 8.1 Example values of the largest acceptable values of χ^2_ν obtained from the $\left(\chi^2_\nu + 3\sigma\right)$ confidence level for different degrees of freedom, ν.

ν	$\left(\chi^2_\nu + 3\sigma\right)$
5	2.9
10	2.3
20	1.9
30	1.8
50	1.6
100	1.4
500	1.2

- For a reasonable fit the value of $\chi^2_\nu \approx 1$.
- If $\chi^2_\nu \ll 1$ check your calculations for the uncertainties in the measurements, α_i.
- The null hypothesis is **questioned** if $\chi^2_\nu > 2$ for $\nu \approx 10$.
- The null hypothesis is **questioned** if $\chi^2_\nu > 1.5$ if ν is in the approximate range $50 \leq \nu \leq 100$.

Although easier to calculate, χ^2_ν does not contain as much information as using χ^2_{min} and ν to calculate $P\left(\chi^2_{min}; \nu\right)$; see Exercise (8.2).

8.4.2   Testing the null hypothesis—a summary

The null hypothesis is that the sample distribution is well modelled by a proposed parent distribution and that any scatter between the two distributions is a result of random statistical variations in the sample. A χ^2 test of this hypothesis is performed by first finding the value of χ^2_{min} and then determining the number of degrees of freedom. We have seen in the previous sections that there are two numbers which can be used to test the validity of the null hypothesis:

(1) the reduced chi-squared statistic, χ^2_ν ;
(2) the probability of obtaining a value of χ^2_{min} equal to the fit value or higher, given ν, $P(\chi^2_{min}; \nu)$.

The χ^2_ν statistic is significantly easier to calculate and can be used to **reject** the null hypothesis if $\chi^2_\nu > 3$. Ambiguity arises in using the χ^2_ν statistic to reject the null hypothesis for values in the range $1 \leq \chi^2_\nu \leq 3$ due to a strong dependence of the χ^2_ν confidence levels on ν as seen in Table 8.1. This ambiguity can be resolved by calculating (using appropriate software or look-up tables) the probability of obtaining the observed value of χ^2_{min} or higher for the number of degrees of freedom, $P(\chi^2_{min}; \nu)$. This probability has a much weaker dependence on ν and the null hypothesis should be rejected if $P(\chi^2_{min}; \nu) < 10^{-4}$.

By rejecting the null hypothesis we are stating that the discrepancies between the data and proposed model are very unlikely to be due to random statistical fluctuations; there must be a genuine systematic reason for the disagreement. A good experimentalist would then try different models with insight gained from further analysis of, for example, the normalised residuals.

8.5   Testing the quality of a fit using χ^2

The χ^2 statistic can be used as a test of the quality of a fit. In this section we discuss by example (a) how to answer the question 'are the data well described by the theoretical model?', and (b) how to answer the related question 'which model best describes the data?'.

The first step in conducting a χ^2 test is to determine the best-fit model function. The procedure to find χ^2_{min} depends on the functional form of the theoretical model. We saw in Chapter 6 that for a simple straight line the best-fit line is easily determined analytically.[4] If the theoretical model is an arbitrary function we highlighted in Chapters 6 and 7 the numerical techniques for minimising χ^2.

The null hypothesis is that the model is an appropriate description of the data. If the null hypothesis is *not rejected* the model is said to be a 'good fit' to the data. If the values of either χ^2_ν or $P(\chi^2_{min}; \nu)$ exceed the limits discussed above, the null hypothesis is rejected and the model is said to be a 'poor fit' to the data. While it is clear what the extremes of a good and poor fit are, the boundary between them is much more subjective. Our advice is to quote both

[4]For a straight-line fit the procedure to obtain χ^2_{min} is as follows:

- obtain the best-fit parameters and their uncertainties using eqns (6.3)–(6.6);
- for each data point, y_i, calculate the normalised residual, $R_i = \frac{y_i - y(x_i)}{\alpha_i}$, using the best-fit function, $y(x_i)$;
- calculate χ^2_{min} by summing the square of the normalised residuals for the data set.

the value of χ^2_{min}, the number of degrees of freedom, ν, and $P(\chi^2_{min}; \nu)$ which allows the reader to judge the quality of the fit.

We now show through some worked examples how the χ^2-test can be applied to testing the quality of a fit and testing different models.

8.5.1 Worked example 1—testing the quality of a fit

In Fig. 8.6 we show the data obtained by monitoring the radioactive decay from an active ^{137}Ba isotope. The count rate is expected to decay exponentially to a constant background level. A suitable model to fit the data will contain three parameters; the initial activity, the half-life of the decay and the background level. Using the procedures outlined in Chapter 6, these parameters and their associated errors were obtained by minimising χ^2 to obtain a value of $\chi^2_{min} = 53.5$. There are 62 data points and as there are three fit parameters the number of degrees of freedom is $\nu = 59$. The reduced χ^2 value is therefore $\chi^2_\nu = 0.9$. As this is less than 1.5 (Table 8.1), and not significantly less than 1, we do not reject the null hypothesis. We can further quantify the quality of fit by using eqn (8.4) to determine the probability of obtaining the value $\chi^2_{min} = 53.5$, or larger, for 59 degrees of freedom. Using suitable look-up tables we find $P(53.5; 59) = 0.68$. As this probability is close to 0.5 there is, again, no reason to reject the null hypothesis and we can accept the fitted parameters of the hypothesised linear function and their uncertainties with a high degree of confidence. In this example we obtain a value for the half-life of the radioactive isotope[5] of $t_{1/2} = 153 \pm 4$ seconds from the best-fit curve.

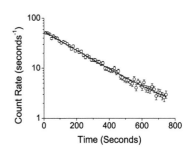

Fig. 8.6 The count rate (the number of decays per second) as a function of time for the decay of a ^{137}Ba isotope. The weighted best-fit curve is shown as a solid line.

[5]The half-life of ^{137}Ba is 153 s.

8.5.2 Worked example 2—testing different models to a data set

As χ^2_{min} is a numerical measure of the discrepancy between a proposed model and the data, weighted by the uncertainty in the data, a χ^2 test can be used to distinguish between the validity of different models. Consider the well-known case for the period of a pendulum. The conventional derivation of the equation of motion for a pendulum assumes that the initial angular displacement is small. If the small-angle approximation is no longer valid, then the period, T, of a pendulum depends not only on its length, L, and the acceleration due to gravity, g, but is also a function of the initial angular displacement, θ_{max}. It can be shown (Kleppner and Kolenkow 1978, p. 276) that including the lowest-order correction yields the result:

$$T = 2\pi \sqrt{\frac{L}{g}} \left[1 + \frac{1}{16} (\theta_{max})^2 \right]. \tag{8.7}$$

In Fig. 8.7 we present experimental data of the period against the initial angular displacement and fit different models to the data.

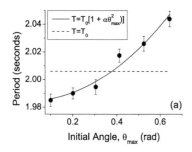

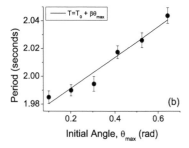

Fig. 8.7 Experimental data (filled circles) for the dependence of the period of a pendulum on the initial angular displacement. In (a) the best-fit constant period is shown as a dashed line, and a model comprising a constant with a quadratic correction shown as the solid line. In (b) a straight-line fit is made to the data.

[6]Further more precise experimental data would need to be obtained to differentiate between the two models. More data would enable a more robust hypothesis test, and the precision of the experimental data enables the subtleties in the shape of the two model functions to be compared with the data.

In Fig. 8.7(a) we fit the data to two competing models motivated by physical insights. We compare the simple case where the period is assumed to have no angular dependence—i.e. the experimental data are fit to a simple constant, $T_0 = 2\pi\sqrt{L/g}$, with a more rigorous approach where the data are fit to eqn (8.7) with a coefficient of the quadratic correction, α, a free parameter: $T = T_0[1 + \alpha\,\theta_{max}^2]$. In Fig. 8.7(b) we test an alternative hypothesis that the angular dependence of the period follows a linear dependence by fitting the data to $T = T_0[1 + \beta\,\theta_{max}]$. Visually the experimental data clearly show an angular dependence and the hypothesis that the period is independent of angular amplitude should be rejected—the best-fit constant period does not agree with any of the six data points within their error bars. To quantify why this null hypothesis should be rejected and whether the other models provide a good fit to the data requires a χ^2 test to be undertaken. The results are encapsulated in Table 8.2.

Table 8.2 Three different models are used for the dependence of the period of oscillation of a pendulum on the initial angular displacement.

Model	Degrees of freedom	χ^2_{min}	χ^2_ν	$P(\chi^2_{min}; \nu)$
$T = T_0$	5	107.2	21.4	1.6×10^{-21}
$T = T_0\left[1 + \alpha\,\theta_{max}^2\right]$	4	3.39	0.9	0.49
$T = T_0[1 + \beta\,\theta_{max}]$	4	4.39	1.1	0.36

The model of a period independent of amplitude can clearly be rejected. With five degrees of freedom a value of $\chi^2_{min} = 107.2$ yields a reduced χ^2_ν of 21.4, and a probability of obtaining a minimum χ^2 of this magnitude or larger of 1×10^{-21}. The model of a quadratic correction to the constant provides a good fit to the data, with values of χ^2_ν and $P(\chi^2_{min}; \nu)$ which are consistent with the model being a valid description of the data. A model with a linear correction also provides a good fit to the data. Based on a χ^2 test alone we cannot distinguish between the quality of the fit for the linear and quadratic models. However, as there is a theoretical model which predicts a quadratic contribution we prefer to accept this model, but note that for this data set we still do not reject the model of a linear correction.[6]

8.5.3   What constitutes a good fit?

Over the last four chapters we have discussed many criteria for deciding whether a theoretical model is an appropriate description of a data set—i.e. 'is the fit good?'. At the start of Chapter 6 we introduced the three questions one should ask when fitting experimental data to a theory. The first, and most important of these, is to question the quality of the fit as this is a prerequisite for the other two questions to be relevant. Often it is clear from a simple visual inspection whether a fit is good or poor, but where this is not the case a more quantitative analysis is required. We provide here a summary of possible strategies to follow in attempting to quantify the 'quality of the fit'. For a good fit

- Two-thirds of the data points should be within one standard error of the theoretical model.
- χ_ν^2 is ≈ 1.
- $P\left(\chi_{\min}^2; \nu\right) \approx 0.5$.
- A visual inspection of the residuals shows no structure.
- A test of the autocorrelation of the normalised residuals yields $\mathcal{D} \approx 2$.
- The histogram of the normalised residuals should be Gaussian, centred on zero, with a standard deviation of 1.

8.6 Testing distributions using χ^2

We have shown how to use the minimisation of the χ^2 statistic to find the best-fit parameters, and use the probability distribution function of χ^2 to perform a hypothesis test. In this section, we shall extend the use of the χ^2 test and demonstrate how to perform a null hypothesis test of a distribution. As before, the null hypothesis is that there is no significant difference between the sample and parent populations—any observed difference is only due to statistical fluctuations associated with the random sampling of the parent distribution.

Suppose that the sample distribution is composed of N measurements of a variable x_i. The first step in testing a sample distribution against a parent distribution is to create a histogram of the sample data. The occurrence per bin in the histogram, O_i, is a function of (1) the sample size and (2) the bin-width. We would expect that if we were to repeat the experiment many times, the number of occurrences per bin would fluctuate about a mean value with a particular standard deviation. As each of these repeat measurements gives a count per bin which is a random process we would expect from Section 3.4 that the appropriate probability distribution function is a Poisson with a mean occurrence O_i, and standard deviation $\sqrt{O_i}$.

The second step is to create the histogram of the expected results. There are two methods for calculating the expected number, E_i, depending on whether the proposed distribution is discrete or continuous. For a discrete distribution E_i is generated by summing the expected occurrences for the range of x-values of the ith bin, and multiplying by N, the total number of measurements. For a continuous probability distribution function $P_{\mathrm{DF}}(x)$, E_i is calculated by integrating $P_{\mathrm{DF}}(x)$ over the range of x-values of the ith bin, and multiplying by N.

In performing a χ^2 test we compare the observed occurrences, O_i, with the expected occurrences, E_i, generated from the proposed parent distribution. The appropriate form for calculating χ^2 is given by eqn (6.2). For the null hypothesis not to be rejected we would expect that $O_i - E_i$ would be small and of the order of $\sqrt{E_i}$. Sequential bins are combined if $E_i < 5$ to avoid the χ^2 test being skewed by the asymmetry in the Poisson distribution.

Once the value of χ^2 has been determined the null hypothesis can be tested by considering the probability $P\left(\chi^2; \nu\right)$ or χ_ν^2 .

The procedure to perform a χ^2 test for a distribution is as follows:

- Construct a histogram of the sample distribution, O_i.
- For the same intervals construct a histogram of the expected occurrences, E_i.
- Combine sequential bins until $E_i > 5$ for all bins.
- Calculate χ^2 using $\chi^2 = \sum_i \dfrac{(O_i - E_i)^2}{E_i}$.
- A good fit will have: $\chi^2_\nu \approx 1$ and $P\left(\chi^2; \nu\right) \approx 0.5$.
- A poor fit will have: $\chi^2_\nu \ll 1$ or $\chi^2_\nu > 3$, and $P\left(\chi^2; \nu\right) \rightarrow 1$ or $P\left(\chi^2; \nu\right) < 10^{-4}$.

In the following section, we give examples of how to test a sample distribution against parent distributions which are (a) discrete and (b) continuous.

8.6.1   Worked example 3—testing a discrete distribution

Consider the data shown in Fig. 8.8 which gives the results of a radioactive decay experiment. From the discussions in Chapter 3 we expect that this sample distribution is drawn from a Poisson distribution. Figure 8.8(a) shows the histogram obtained after taking 58 measurements of one second duration. Our best estimate of the mean of the parent distribution is the mean number of counts determined from the sample distribution—in this case $\bar{N} = 7.55$. The expected values for a Poisson distribution with this mean are superimposed on the histogram in Fig. 8.8(a). To calculate the expected values, the probabilities determined from the Poisson distribution function are scaled by the total number of samples, N; $E_i = P\left(N_i, \overline{N}\right) \times N$.

Visually it appears that the sample distribution is in agreement with a Poisson distribution but for a quantitative analysis we calculate χ^2; these calculations are carried out explicitly in Table 8.3.

Due to the low number of expected occurrences in bins 0–4 and 11–17, these have been combined to form eight super-bins which are plotted in Fig. 8.8 (b). A visual inspection of the re-binned histogram reveals that for six of the eight bins the observed and expected occurrences agree within the expected Poisson fluctuations, $\sqrt{E_i}$. This is consistent with a good fit, but a more quantitative analysis of the null hypothesis requires χ^2 to be calculated. We have applied two constraints when comparing the sample and parent distributions—namely the means of the two distributions are the same, and the total number of measurements is also the same. The number of degrees of freedom in this example is therefore $\nu = 8 - 2 = 6$. We determine $\chi^2 = 6.80$ in Table 8.3 and calculate both the reduced chi-squared, $\chi^2_\nu = 1.13$, and the probability $P(6.80; 6) = 0.34$. As χ^2_ν is close to 1 and the probability is close to 0.5, there is no reason why the null hypothesis should be rejected and we conclude that the sample distribution was likely to have been drawn from a Poisson parent distribution.

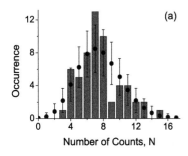

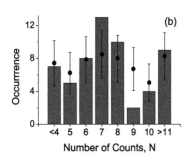

Fig. 8.8 Comparing experimental radioactive decay events with a Poisson distribution. In (a) a histogram of the observed occurrence of counts is plotted (bars) and compared with a Poisson distribution (points) and expected fluctuations. Part (b) shows a re-binned histogram with bin-widths chosen such that the number of expected occurrences is always greater than 5. Note that in the re-binned histogram six out of eight observed occurrences fall within the expected range taking into account the Poisson noise.

Table 8.3 Comparing experimental radioactive decays with a Poisson model. The first column is the number of counts, the second is the occurrence of each count. Column three is the Poisson probability for obtaining a given number of counts using the mean count of the data. The fourth column gives the expected number of occurrences, and χ^2 is calculated in the fifth column.

Number of counts	O_i		Prob. (%)	E_i		$\dfrac{(O_i - E_i)^2}{E_i}$
0	0 ⎫		0.05	0.03 ⎫		
1	0 ⎪		0.40	0.23 ⎪		
2	0 ⎬ 7		1.50	0.87 ⎬ 7.44		0.026
3	1 ⎪		3.77	2.19 ⎪		
4	6 ⎭		7.12	4.13 ⎭		
5	5		10.75	6.23		0.24
6	8		13.53	7.85		0.003
7	13		14.60	8.47		2.43
8	10		13.78	7.99		0.51
9	2		11.56	6.71		3.30
10	4		8.73	5.06		0.22
11	4 ⎫		5.99	3.48 ⎫		
12	2 ⎪		3.77	2.19 ⎪		
13	2 ⎪		2.19	1.27 ⎪		
14	0 ⎬ 9		1.18	0.69 ⎬ 8.25		0.07
15	1 ⎪		0.60	0.35 ⎪		
16	0 ⎪		0.28	0.16 ⎪		
> 17	0 ⎭		0.21	0.12 ⎭		
	$\sum 58$		$\sum 100$	$\sum 58$		$\chi^2 = \sum 6.80$

8.6.2 Worked example 4—testing a continuous distribution

In the discussion of the central limit theorem in Chapter 3 the key idea is that, independent of the form of the distribution function of individual measurements, the distribution of the means of, say, five measurements is Gaussian. As an example we showed the distribution of the average of the six balls drawn in the UK lottery for 106 independent events. A visual inspection of Fig. 3.9(b) indicates that the histogram of the means resembles a Gaussian. Here we provide a qualitative analysis of this agreement through a χ^2 test. The histogram of the means is reproduced in Fig. 8.9. The proposed continuous Gaussian model with the same mean ($\bar{x} = 25.4$) and standard deviation ($\sigma = 5.8$) as the sample and scaled by the total number of sample occurrences, $N = 106$, is shown by the solid line in Fig. 8.9. To compare the histogram of the observed means with the continuous function requires the Gaussian to be discretised to the same bin-width as the histogram of the observed occurrences. The expected counts per bin were calculated using eqn (3.9) to determine the probability and multiplying by N. In Table 8.4 the data have been re-binned to ensure that E_i is always larger than 5 in any bin resulting in a new sample of 16 measurements. As there are three constraints (the parent and sample distributions have the same number of measurements, and share a common mean and standard deviation) the number of degrees of freedom is therefore $\nu = 16 - 3 = 13$. We determine

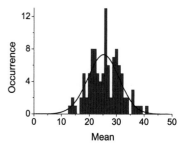

Fig. 8.9 A histogram of the means of 106 UK National Lottery draws. In each draw six balls are selected from the integers 1–49 and we have calculated the average value of the balls drawn per event. The continuous line is a Gaussian constrained to have the same mean and standard deviation as the sample data and is scaled by the total number of events.

Table 8.4 Comparing the distribution of the means of the 106 UK National Lottery draws with a Gaussian distribution. The first column is the mean value, the second is the occurrence of each mean. Column three gives the expected number of occurrences using a Gaussian model with the same mean and standard deviation as the sample data, and χ^2 is calculated in the final column. Bin-widths have been chosen such that E_i is greater than 5 in each bin. Note that all the entries in the last column are of order 1.

Bin	O_i	E_i	$\dfrac{(O_i - E_i)^2}{E_i}$
<16	4	6.3	0.85
17–18	7	5.6	0.33
19–20	7	8.5	0.28
21	8	5.4	1.25
22	8	6.1	0.60
23	5	6.7	0.41
24	4	7.1	1.33
25	5	7.3	0.72
26	13	7.3	4.48
27	6	7.1	0.16
28	2	6.7	3.26
29	7	6.1	0.14
30	8	5.4	1.25
31–32	11	8.5	0.71
33–34	4	5.6	0.47
>35	7	6.3	0.08
$\sum$	106	106	16.32

[7]Named after William of Occam, a fourteenth century Franciscan friar, who postulated that with two competing theories which make the same prediction, the simpler one is better.

Fig. 8.10 Different polynomial fits to a data set with 41 points. Low-order polynomial fits such as the second-order fit shown in (a) systematically fail to account for the trends in the data set. Higher order polynomials fit the data set better. The evolution in the quality of fit, χ^2, is shown as a function of polynomial order in (b). Fits with polynomial orders greater than five do not significantly improve the quality of the fit.

that $\chi^2 = 16.32$ in Table 8.4 and calculate both the reduced chi-squared, $\chi_\nu^2 = 1.26$, and the probability $P\,(16.32;\,13) = 0.23$. As χ_ν^2 is close to 1 and the probability is close to 0.5, there is no reason why the null hypothesis should be rejected and we conclude that the distribution of means is well modelled by a Gaussian distribution in agreement with the central limit theorem.

One can adopt this methodology to provide a quantitative test as to whether the histogram of the normalised residuals obtained in a fit is Gaussian as expected.

8.7   Occam's razor

It should be obvious that by adding more parameters to our model, we can get a better fit to the experimental data points. As mentioned above in the discussion of the number of degrees of freedom, a fit is much more convincing if there are far fewer constraints than data points. When we have fewer constraints than data points it is useful to consider whether the model would be better with more (or fewer) parameters. We will use a line of argument based on a tool designated 'Occam's razor' to ascertain whether another parameter is justified in the theoretical model.[7]

Consider the data shown in Fig. 8.10. In part (a) a polynomial fit is hypothesised as a theoretical model, and the value of $\chi_{\min}^2$ is plotted as a function of the number of parameters included in Fig. 8.10(b). We see that the quality of the fit improves drastically up until the fifth-order term is included. For higher order polynomials there is only a modest improvement in the quality of the fit as more terms are added, reflected in the slight decrease of $\chi_{\min}^2$. As the quality of the fit hardly improves after the fifth-order term is included, we apply Occam's razor to discard higher order corrections, and prefer to accept the simplest model which accounts quantitatively for the trends in the data. It is also possible to apply a more quantitative test of whether an additional term should be kept, in terms of an F-test; see Bevington and Robinson (2003, Section 11.4) for further details.

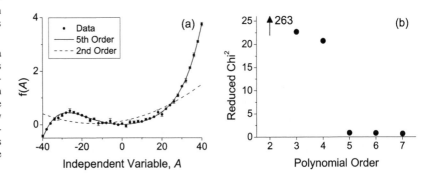

8.8 Student's *t*-distribution

In Chapter 3 we discussed how to compare experimental results with an accepted value. The analysis centred on the dimensionless quantity

$$t = \frac{(\bar{x} - \mu)}{\alpha} = \frac{\sqrt{N}\,(\bar{x} - \mu)}{\sigma_{N-1}}, \qquad (8.8)$$

where $\bar{x}$ is the best estimate based on N measurements with sample standard deviation of σ_{N-1}, standard error α, and accepted value μ. Calculations of the confidence limits were conducted with this parameter; for example we showed that 99% of data points for a Gaussian distribution should have $|t| < 2.58$, 95% have $|t| < 1.96$ and 68% have $|t| < 1$. However, in practice we do not know the standard deviation of the parent distribution and we can only estimate it from the sample distribution. For samples of finite size the confidence limit becomes a function both of the tolerance level chosen and the number of degrees of freedom, ν. The factors which replace the entries in Table 3.1 are known as the Student *t* values, and are derived from a well-known distribution.[8] When the number of degrees of freedom is large, the distribution approximates well to a Gaussian; but for fewer degrees of freedom the Student *t* distribution is wider than a Gaussian. Most spreadsheets and analysis packages can calculate the relevant factor from the desired confidence limit and the number of degrees of freedom. The evolution of the factor for the 68%, 95% and 99% confidence limits is shown in Fig. 8.11. The difference between the confidence limits derived from the Student and Gaussian distributions depends on both the confidence limit of interest and the number of degrees of freedom. For the 68% confidence limit the difference is only 5% for 10 degrees of freedom and 10% for five degrees of freedom. As we seldom quote the uncertainties to more than one significant figure we do not have to worry about this effect unless the number of degrees of freedom is very small. The Student probability distribution function has higher cummulative probabilities for large deviations than a Gaussian. Thus, the importance of the *t* values increases as the confidence level tends to 100%.

[8]W. S. Gosset published the distribution using the pseudonym 'Student' in 1906 while an employee of the Guinness brewery.

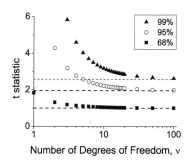

Fig. 8.11 The variation of the *t* statistic as function of the number of degrees of freedom for the 68%, 95% and 99% confidence limits. As $\nu \to \infty$, the values tend to those obtained using a Gaussian distribution (1, 1.96 and 2.58 respectively). For very few degrees of freedom the confidence limits have to be broadened significantly.

8.9 Scaling uncertainties

We have emphasised throughout this book that it is vital to ascertain the magnitude of the uncertainty in the measurements, α_i, and have shown how it is possible to extract uncertainties in parameters from an analysis of the goodness of fit. It is also possible to turn this process on its head, and learn something about the uncertainties from the fit. There are two separate procedures which we discuss here: (i) estimating the **common uncertainty** on the data points, and (ii) scaling the magnitude of the uncertainties in fit parameters.

Both processes hinge on the concept that a 'good fit' will have a $\chi^2_{\min} \approx \nu$. Recalling the definition of the standard error, α_i, as the standard deviation of

[9]**Health warning** Do not use the data and fit parameters to estimate the uncertainty, and then use the common uncertainty to calculate χ^2 and hence test the quality of fit—this is a circular argument.

the mean, one would expect the statistical variation between the theory and the data, $y_i - y(x_i)$, to be of the order of the standard error. One can estimate the common uncertainty, α_{CU}, in the measurements by setting $\chi^2_\nu = 1$. If the data set is homoscedastic one can use eqn (6.1), to give

$$\alpha^2_{CU} = \frac{1}{\nu} \sum (y_i - y(x_i))^2. \tag{8.9}$$

In most statistical regression packages the common error is returned. Note that one cannot use α_{CU} to give any insight into the quality of the fit.[9]

It is also possible to use the value of χ^2_{min} to scale the errors in the fit parameters when one has complete confidence in the theoretical model being used to describe the experimental data. Let S be the **scale factor** defined as

$$S = \sqrt{\frac{\chi^2_{min}}{\nu}} = \sqrt{\chi^2_\nu}. \tag{8.10}$$

One can rewrite the discussion from Section 8.4.1 in terms of S. Thus, if the results of a χ^2 minimisation yield a value of S that is very large, the model should be rejected and the experimenter should try and ascertain the reason behind the large discrepancy between theory and experiment. Similarly the parameters should be rejected if S is very small as it is likely that the error bars have been overestimated. If S is similar to 1 the uncertainties in the fitted parameters can be scaled by S. The reasoning is (see page 16 of the article by the Particle Data Group, Amsler *et al.* 2008) that the deviation of S from 1 is a consequence of the uncertainties in the experimental data being estimated incorrectly. Not knowing which particular points give rise to the unexpected value of S, it is assumed that all the uncertainties are incorrectly scaled by a common factor of S. By scaling the uncertainties in the input data by S, the uncertainties in the fit parameters also change by S giving a modified value of $\chi^2_\nu = 1$ (this is equivalent to scaling every element of the error matrix by S). Obviously there are many assumptions inherent in this scaling procedure, and the experimenter has to ask whether they are reasonable and justified. If a scaling factor is applied to the uncertainties, this should be clearly stated in any publication based on these results.[10]

[10]**Health warning** Note that many fitting packages automatically scale the fit parameters, often without making it clear how the data have been manipulated. Many packages will blindly apply the scaling even if $\chi^2_\nu \ll 1$ due to incorrect error analysis, leading to completely unrealistic estimates of the uncertainties.

8.10 Summary of fitting experimental data to a theoretical model

In the last four chapters we have discussed extensively various issues which arise when fitting experimental data to a theoretical model. Here we summarise the procedure and highlight the questions one should ask after performing such a fit.

- Perform the fit and find optimal values of the parameters by minimising χ^2.
- Based on the value of χ^2_{min} and ν decide whether the fit is reasonable—i.e. questions such as 'are my data consistent with a Gaussian distribution?' should be answered at this stage.
- If there are competing theoretical models use χ^2_{min} and ν to decide which model is most appropriate.
- If the quality of the fit is poor either (a) consider a different theoretical model, or if the theoretical model is known to be valid for the conditions of the experiment, (b) try to identify defects in the experiment or analysis.
- For a good fit the values of the parameters which minimised χ^2 are used to answer the question 'what are the best-fit parameters?'.
- Calculate the error matrix and use the square root of the diagonal elements to answer the question 'what are the uncertainties in the best-fit parameters?'.

Chapter summary

- The χ^2 statistic can be used for hypothesis testing.
- The question 'are my data consistent with the proposed model?' can be answered by analysing the value of χ^2_{min}.
- The number of degrees of freedom, ν, is equal to the number of data points, N, less the number of constraints, $\mathcal{N}$.
- For a good fit, the expectation value of χ^2 is ν, with a standard deviation of $\sqrt{2\nu}$.
- If $\chi^2_{min} > \nu + 3\sqrt{2\nu}$ the null hypothesis is rejected.
- Reduced chi-squared, χ^2_ν, is defined as $\chi^2_\nu = \dfrac{\chi^2_{min}}{\nu}$.
- Occam's razor can be used to eliminate unwarranted extra parameters in a theoretical model.
- Student's t distribution should be used when comparing experimental results with an accepted value with a small number of degrees of freedom.
- It is possible to scale error bars and estimate the statistical fluctuations in a data set if one is confident that the appropriate theoretical model has been used to describe the data (at the expense of foregoing any discussion about the quality of the fit).

Exercises

(8.1) *Confidence limits for χ_ν^2*

Make a table similar to Table 8.1 and calculate the 68% and 90% confidence limits as a function of the degrees of freedom for the χ_ν^2 statistic.

(8.2) *Confidence limits and χ_ν^2*

After performing a fit with 10 degrees of freedom the value of χ_{min}^2 is found to be 15.9. Calculate (a) the value of χ_ν^2, and (b) the probability of obtaining a value of χ_{min}^2 equal to this value or larger given the degrees of freedom. In another fit with 100 degrees of freedom, the value of χ_{min}^2 is 159. Calculate (c) the value of χ_ν^2, and (d) the probability of obtaining a value of χ_{min}^2 equal to this value or larger given the degrees of freedom. Comment on the differences between the values obtained in part (b) and (d).

(8.3) *Does the noise on a photodiode signal follow a Gaussian distribution?*

As we discussed in Chapter 1, for very low intensities the distribution of counts from a photodetector is expected to follow the Poisson shot-noise distribution. However, for larger photon fluxes the noise on the voltage generated in a photodiode circuit is expected to follow a Gaussian distribution. Figure 3.4 shows the signal output from a photodiode as a function of time, and in part (b) a histogram of the distribution of data. The number of observed data points lying within specified bands, O_i, is given below.

Interval/σ	$(-\infty, -2.5)$	$(-2.5, -2)$	$(-2, -1.5)$
O	9	48	142
Interval/σ	$(-1.5, -1)$	$(-1, -0.5)$	$(-0.5, 0)$
O	154	438	521
Interval/σ	$(0, 0.5)$	$(0.5, 1)$	$(1, 1.5)$
O	405	318	299
Interval/σ	$(1.5, 2)$	$(2, 2.5)$	$(2.5, \infty)$
O	100	57	9

(i) Use eqn (3.9) to determine the number of data points expected in each interval, E_i. (ii) Show that $E_i > 5$ for all bins, and hence there is no need to combine

sequential bins. (iii) Calculate χ^2 from the formula $\chi^2 = \sum_{i=1}^{12} (O_i - E_i)^2/E_i$. (iv) Calculate the number of degrees of freedom. (v) Are the data consistent with the hypothesis of a Gaussian distribution?

(8.4) *Is the distribution of occurrences of balls in the National Lottery uniform?*

In Fig. 3.9 we showed the histogram of the occurrences of the 49 balls in all 106 National Lottery draws for the year 2000. The data are reproduced below. Test the hypothesis that the balls are chosen at random, i.e. that the distribution of occurrences is uniform.

N	1	2	3	4	5	6	7	8	9	10
O	11	11	13	14	11	22	15	9	9	16
N	11	12	13	14	15	16	17	18	19	20
O	17	12	8	13	8	15	9	13	19	9
N	21	22	23	24	25	26	27	28	29	30
O	12	10	17	13	10	9	10	15	9	14
N	31	32	33	34	35	36	37	38	39	40
O	16	17	11	13	14	11	13	21	14	13
N	41	42	43	44	45	46	47	48	49	
O	12	11	16	13	10	18	16	16	8	

(i) Recalling that six balls are selected per draw, and assuming a uniform distribution, calculate the expected number of occurrences of each ball, E_i. (ii) Show that $E_i > 5$ for all bins, and hence that there is no need to combine sequential bins. (iii) Calculate χ^2 from the formula $\chi^2 = \sum_{i=1}^{49} (O_i - E_i)^2/E_i$. (iv) Calculate the number of degrees of freedom. (v) Are the data consistent with the hypothesis of a uniform distribution of occurrences?

(8.5) *Is the temporal distribution of goals in a football game uniform?*

In September 2009, 101 goals were scored in the English Premier League. A breakdown of the observed number of goals, O_i, during nine equal-duration time intervals is given below. Test the hypothesis that there is no preferential time during the game at which goals are scored, i.e. that the time distribution of goals is uniform.

Interval/minutes	(1,10)	(11, 20)	(21, 30)
Goals scored	6	11	8
Interval/minutes	(31, 40)	(41, 50)	(51, 60)
Goals scored	8	14	12
Interval/minutes	(61, 70)	(71, 80)	(81, 90)
Goals scored	11	12	19

(i) Assuming a uniform distribution, calculate the expected number of goals per interval, E_i. (ii) Show that $E_i > 5$ for all bins, and hence there is no need to combine sequential bins. (iii) Calculate χ^2 from the formula $\chi^2 = \sum_{i=1}^{9} (O_i - E_i)^2/E_i$. (iv) Calculate the number of degrees of freedom. (v) Are the data consistent with the hypothesis of a uniform distribution?

(8.6) *Does the distribution of goals per side per game follow a Poisson distribution?*

In the first six weeks of the 2009 English Premier League season, 229 goals were scored in the 76 games. Does the number of goals per game per side follow a Poisson distribution? A breakdown of the occurrence, O_i, of games in which N goals were scored by a side is given below.

Goals/side/game	0	1	2	3	4	5	6
Number of games	38	51	35	14	7	4	3

(i) Calculate the mean number of goals per side per game, $\overline{N}$. (ii) Assuming a Poisson distribution, calculate the expected number of goals per side per game, E_i. (iii) Ascertain whether some of the bins should be combined. (iv) Calculate χ^2 and the number of degrees of freedom. (v) Test the hypothesis that the number of goals per side per game follows a Poisson distribution.

(8.7) *Is a die fair?*

A die was thrown 100 times, and the number of times each face landed up is given below.

Face value	1	2	3	4	5	6
O	17	21	14	13	16	19

If the die is fair the expected number of occurrences would be the same for each number. Test the hypothesis that the die is fair.

(8.8) *Is a straight line a good fit to the data?*

Use the data tabulated in Exercises (6.5) and (7.8) and the best-fit straight line intercepts and gradients to calculate χ^2 for the fits. How many degrees of freedom are there for these fits? Is it fair to conclude that the data are well fit by a straight line for these cases?

Topics for further study

9

In this chapter we briefly discuss some topics which had to be omitted from this book for the sake of brevity, and give an indication of where further details might be obtained. A recurring theme in this chapter will be that owing to the computational power afforded by modern computers it is often far easier to generate and analyse synthetic data than it is to look for analytic solutions to difficult problems.

9.1 Least-squares fitting with uncertainties in both variables

We have used the method of least squares extensively throughout this book to obtain best fits of a model to a data set, subject to the assumption that only the uncertainty in the dependent variable, y, was significant. There are many instances when the uncertainty in the independent variable, x, is also significant, and should be included in the calculations. Here we briefly discuss some of the issues which arise and possible strategies for including the uncertainties in both variables into the analysis.

9.1.1 Fitting to a straight line

For the straight line $y = mx + c$, we previously included the y-uncertainty, α_{yi}, in our analysis. If there is also an x-uncertainty, α_{xi}, this can be viewed as generating an equivalent y-uncertainty of $m\,\alpha_{xi}$, as is evident from Fig. 9.1. Therefore we can modify the definition of χ^2 to be:

$$\chi^2(m, c) = \sum_{i=1}^{N} \frac{(y_i - mx_i - c)^2}{\alpha_{yi}^2 + m^2 \alpha_{xi}^2}. \tag{9.1}$$

The weight, w_i, is defined as

$$\frac{1}{w_i} = \alpha_{yi}^2 + m^2 \alpha_{xi}^2, \tag{9.2}$$

and can be interpreted as the inverse of the variance of the linear combination $y_i - mx_i - c$. Equation (9.1) is of the appropriate form, i.e. a sum of N random variables normalised by their variance, for the applicable distribution to be

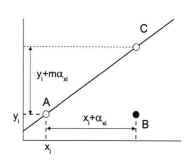

Fig. 9.1 In linear regression, an x-uncertainty, α_{xi}, in the independent variable brings about an equivalent y-uncertainty of $m\,\alpha_{xi}$.

that of χ^2. The difficulty in using eqn (9.1) is the nonlinear dependence of χ^2 on m. Possible strategies to be pursued with this approach are outlined in Press *et al.* (1992, Section 15.3). Finding the uncertainties in m and c is significantly less straightforward than the cases considered in Chapter 7 owing to this nonlinear dependence of χ^2 on m. If the errors follow a Gaussian distribution then numerical techniques are used to locate the $\Delta\chi^2 = 1$ contour (Press *et al.* 1992, Section 15.3); if the errors are not normally distributed a Monte Carlo technique is usually adopted to quantify the uncertainties (see Section 9.3).

9.1.2 Fitting to a more general function

For the general case we can define a sum of the weighted squared residuals, S, defined as:

$$S = \sum_{i=1}^{N} \left[w_{x\,i} (x_i - X_i)^2 + w_{y\,i} (y_i - Y_i)^2 \right]. \tag{9.3}$$

Here x_i and y_i are the experimental measurements of the two variables, and X_i and Y_i are predicted, or calculated, values subject to some model. Typically, the weightings are chosen to be the inverse of the variances, $w_{x\,i} = 1/\alpha_{x\,i}^2$, and $w_{y\,i} = 1/\alpha_{y\,i}^2$, but other models are also used. Therefore the sum does not, in general, have the χ^2 distribution of Chapter 8, which is why we use another symbol. The papers by Reed (1989, 1992) and Macdonald and Thompson (1992) discuss this methodology both in the context of fitting a straight line, and more general functions. Macdonald and Thompson's work is an excellent review of previous presentations, and provides a survey of the algorithms adopted to tackle the problem of how to minimise S.

9.1.3 Orthogonal distance regression

Another generalised least-squares method is that of **orthogonal distance regression**. The *modus operandi* of conventional linear regression is to minimise the sum of the squared *vertical distances* (see Fig. 5.7) between the data and the y-coordinates of the fit line for the same x-coordinate. In contrast, in orthogonal distance regression it is the orthogonal distances between the data and the fit line which are minimised, as depicted in Fig. 9.2. Consider once more the straight line $y = mx + c$. An orthogonal line will have a gradient of $-1/m$, and be described by the equation $y' = -x/m + c'$. If this second line passes through the data point (x_1, y_1) its equation becomes $y' = (x_1 - x)/m + y_1$. These two lines will intersect at the point $(x_{\text{Int}}, y_{\text{Int}})$, where the coordinates are given by:

$$x_{\text{Int}} = \frac{x_1 + my_1 - mc}{1 + m^2}, \tag{9.4}$$

$$y_{\text{Int}} = mx_{\text{Int}} + c. \tag{9.5}$$

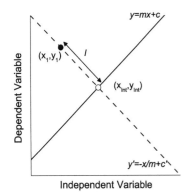

Fig. 9.2 The orthogonal distance, ℓ, of a point from the best-fit straight line is shown. In orthogonal distance regression the sum of the squares of the orthogonal distances for all data points is minimised in order to find the best-fit straight line. The concept can be extended to more general functions.

The orthogonal distance, ℓ, between the data point (x_1, y_1) and the fit line is obtained from the equation:

$$\ell^2 = (x_1 - x_{\text{Int}})^2 + (y_1 - y_{\text{Int}})^2 . \qquad (9.6)$$

In the method of orthogonal distance regression, the sum of the orthogonal distances is minimised by varying the fit parameters (in this case, m and c). We recognise the sum of the squares of the orthogonal distances as a special case of eqn (9.3). The concept can be extended to more general functions, and is clearly described in the paper by Boggs *et al.* (1987). A range of commercial software packages exist for implementing different orthogonal distance regression algorithms.

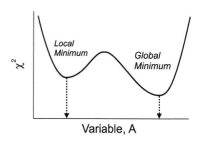

Fig. 9.3 An error surface with more that one minimum. The strategies discussed in earlier chapters are prone to become trapped in the local minimum. Algorithms which incorporate random changes to the search parameters are better at finding the global minimum.

9.2 More complex error surfaces

In Chapters 6–8 we saw numerous examples of error surfaces. All of these had the feature that there was a single well-defined minimum, and we discussed various strategies of how to locate the best parameters. Error surfaces can be more complicated, as shown schematically in Fig. 9.3. We must distinguish between **local minima** and **global minima**. The bane of all of the efficient methods for finding minima we have discussed previously is that they can get 'trapped' at local minima. To overcome this issue there exist different search algorithms which incorporate *random* changes to the search parameters. Often this leads to an increase in the value of χ^2 (or whichever quantity is being optimised), which is to the detriment of the minimisation procedure. However, allowing the option of 'uphill' trajectories across the error surface allows the possibility of escape from local minima. We discuss in slightly more detail two methods which are implemented extensively.

9.2.1 Simulated annealing

This method is based on an insight from statistical mechanics, and is outlined in the article by Kirkpatrick *et al.* (1983). Consider the process by which a metal cools and anneals. At a high temperature the atoms in the metal are free to move, and as the metal is cooled the mobility decreases. The high initial heat allows the atoms to depart from their initial positions, such that, on cooling, they can arrange themselves into a pure crystal with long-range order. The crystalline state represents a lower energy configuration for the system. The process of cooling the metal slowly gives the atoms a higher probability of finding configurations with an internal energy lower than the starting condition. The method hinges on the cooling being sufficiently slow, allowing the atoms to redistribute themselves into a lower-energy configuration as their mobility decreases.

The analogy with finding the global minimum of an error surface is the following. The trajectory across the error surface is driven by updating the values of the fit parameters with a random increment. This corresponds to the role heat plays in the thermodynamic system. The role of 'temperature' is played by a parameter that dictates the probability of the increment being a certain size. The 'temperature' is then slowly reduced to zero when the final values of the parameters chosen are 'frozen in'. For the initial high 'temperature' the parameters essentially change randomly, in analogy with atoms in the metal being very mobile. As the 'temperature' is reduced the trajectory tends to be downhill on the error surface, in analogy with the thermodynamic system finding a lower energy configuration. Crucially the opportunity for uphill motion at finite 'temperature' offers a route out of local minima.

9.2.2   Genetic algorithms

Genetic algorithm techniques, like simulated annealing, can also find the global minimum on the error surface. The first step is to **encode** potential solutions to the problem (for us, the set of parameters which minimise χ^2). This can be done as a bit string[1] in an object known as a **chromosome**, to emphasise the link with evolutionary biology. The algorithm produces **generations** thus:

(1) A large population of *random* chromosomes is generated. Each of these chromosomes encodes the information from different solutions (for our case, these are different values of the parameters).

(2) Each chromosome is assigned a **fitness score**, which is a number to represent how well that particular trial solution solves the problem (the fitness score needs to increase as χ^2 decreases).

(3) Two members of the present population are selected such that a new generation can be bred. The selection procedure is such that chromosomes with a higher fitness score are more likely to be selected.

(4) A 'child' (or new trial solution) is generated by the two 'parents' reproducing. Two genetic operations, **crossover** and **mutation**, are used to produce the next generation. The analogy with biological evolution is that the child shares many of the attributes of its parents. Dependent on the crossover rate (which is set in the algorithm), a bit is randomly chosen in one chromosome, and all subsequent bits in the first chromosome are replaced by the bits from the second chromosome.

(5) Each bit in the selected chromosomes has a random (very small) probability of being flipped (i.e. 0 becomes 1, and vice versa)—this is mutation.

(6) Steps 2–5 are repeated until a new population has been generated.

A consequence of the process utilising selection based on fitness, and reproduction with crossover and mutation, is that the next generation of chromosomes is (i) different from the previous generation, and (ii) typically fitter. New generations are generated until the best solution is obtained, within some tolerance. Like simulated annealing, the stochastic nature of the algorithm enables trial solutions to avoid becoming 'stagnated' in local minima, and the global minimum to be found. Many of the ideas implemented in contemporary genetic algorithms were introduced by Holland (1975); the book contains further details of how to implement the procedure.

[1] 10010010 is an example of an 8-bit binary string.

9.3 Monte Carlo methods

9.3.1 Introduction to Monte Carlo methods

Monte Carlo[2] methods are numerical techniques for calculating quantities such as integrals, probabilities, and confidence limits, by the use of a sequence of random numbers.[3] At one level the ethos of Monte Carlo methods can be viewed as being experimental statistics, i.e. the analytic approach adopted previously is replaced with a numerical methodology. The computational power of modern computers enables a large number of calculations to be performed, thus circumventing in some circumstances a lack of deep theoretical understanding.

We illustrate the methodology by showing the Monte Carlo calculation of π. Consider Fig. 9.4, which shows a circle of diameter 1, circumscribed by a square with sides of length 1. Elementary geometry reveals that the ratio of the area of the circle to the area of the square is $\pi/4$. Now imagine we can generate a random number, r, constrained to the interval $0 < r < 1$. The first two numbers chosen are used as the (x, y) coordinates of a point within the perimeter of the square of Fig. 9.4. Many such pairs are generated, and some of the points generated are depicted in the figure. For each point the following question has to be answered: does the point lie within the circle? After N_{TOT} trials let N_{IN} denote the number of points inside the circle. Our estimate of π is then $\pi = \dfrac{4 \times N_{\text{IN}}}{N_{\text{TOT}}}$. Table 9.1 shows the results obtained as a function of the total number of trials. How does one estimate the uncertainty in the value deduced from a Monte Carlo simulation? Using the relevant distribution for the number of points inside the circle it is expected that the fractional uncertainty in the mean scales as $1/\sqrt{N_{\text{TOT}}}$. There are significantly more efficient methods for calculating π; the purpose of this example was to highlight the Monte Carlo methodology.

[2]The methods are named after the area of Monaco where a famous casino is located. Random numbers are a defining feature of Monte Carlo calculations, much as the laws of chance govern gambling games.

[3]Technically, computers generate pseudo-random numbers, but the difference will not be of concern for us.

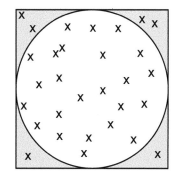

Fig. 9.4 A Monte Carlo method for evaluating π. The coordinates of the points are selected at random. With N_{IN} points inside the circle after N_{TOT} trials, we get the estimate $\pi = 4 \times N_{\text{IN}}/N_{\text{TOT}}$.

9.3.2 Testing distributions with Monte Carlo methods

The example of estimating π was a special case where the random numbers generated in the interval $0 < r < 1$ could be used without having to be processed further. Typically, this is not the case. The more general case is where we are interested in the properties of some probability distribution function, $P_{\text{DF}}(x)$. The **transformational method** is used such that the numbers chosen at random are distributed according to our desired distribution, $P_{\text{DF}}(x)$ (see Cowan 1988 and Bevington and Robinson 2003 for details). Having obtained the appropriate distribution we can proceed to collect synthetic data sets (much as we would collect genuine experimental data from a real experiment). Having obtained a sequence of Monte Carlo generated values we can apply the statistical techniques discussed in earlier chapters to estimate parameters of interest, such as the mean, variance, etc. In common with the specific case of estimating π demonstrated in the previous section, the accuracy of the results scale as $1/\sqrt{N_{\text{TOT}}}$, with N_{TOT} being limited only by the computational power and the complexity of the problem.

Table 9.1 The evolution of the Monte Carlo estimate of π with sample size N_{TOT}. Five different simulations were run for each value of N_{TOT}, from which the mean and uncertainty were calculated.

N_{TOT}	π	Uncertainty
100	3.18	0.05
500	3.15	0.03
1000	3.14	0.03
5000	3.124	0.009
10000	3.145	0.003

One of the most powerful applications of Monte Carlo techniques is to generate confidence limits and distribution functions of random variables. The discussion of the confidence limits obtained from fit parameters in Chapter 7 assumed that the appropriate distribution function was a Gaussian, largely motivated by the central limit theorem. However, there are many examples, especially with nonlinear least-squares fitting, which yield non-normal error distributions. Monte Carlo simulations offer a simple and fast way to map out the distribution function and to ascertain, say, the relevant $\Delta \chi^2$ contour for the confidence limit of interest. The article by Silverman *et al.* (2004) has examples of computer simulated histograms for probability distribution functions of products and quotients of independent random variables. Further details about Monte Carlo simulations can be found in the books by Bevington and Robinson (2003, Chapter 5), Cowan (1998, Chapter 3) and Press *et al.* (1992).

9.4   Bootstrap methods

Often real data sets have probability distributions which do not perfectly match one of the simple, classical distributions we have discussed extensively in the earlier chapters of this book. In this case it is not possible to derive simple analytic results for the confidence limits. Moreover, there are occasions when we want to consider more complicated statistics than the mean, standard deviation, etc. Furthermore, in addition to the question 'what are the confidence limits?', we can also ask 'are these values realistic?'. Efron developed the bootstrap method as a numerical technique that lets us derive confidence intervals for any statistic, on a data set with any probability distribution function. Indeed, the bootstrap method works even if we don't know the probability distribution. The article by Efron and Tibshirani (1986) provides a survey of bootstrap techniques, with the definition of the bootstrap as 'a computer-based method, which substitutes considerable amounts of computation in place of theoretical analysis' (Efron and Tibshirani 1986, p. 54, opening paragraph). Further details of how to implement bootstrap algorithms can be found in Press *et al.* (1992, Section 15.6).

The concept of **replacement** is important for distinguishing among various sampling schemes. Sampling schemes may or may not use replacement—an element from the original set can be selected many times with replacement, but cannot be selected more than once without replacement. For example, consider the case when there are 10 red and 10 blue balls in a bag, and a sample of two is desired. Say we first choose a red ball. With replacement, the first ball drawn is put back into the bag, such that the probability of the second ball drawn being red is one-half, as is the the probability of the second ball drawn being blue. Without replacement, the probability of the second ball drawn being red is $9/19$, and the probability of the second ball drawn being blue is $10/19$.

Consider the case where there are N data points, each of which could either be a single measurement, or a more complex object such as (x_i, y_i) pairs. From these data points a statistic of interest can be calculated (such as the mean, or gradient of the best-fit line). The bootstrap method is implemented as follows:

(1) Generate a **synthetic data set** with the same number of points as the original set by selecting, at random, N data points from the original set, with *replacement*.[4]

(2) Calculate the statistic of interest for the synthetic data set.

(3) Repeat the first two steps a large number of times (typically many hundreds or thousands of times).

(4) The distribution of the large number of computed values of the statistic form an estimate of the sampling distribution of the statistic.

[4]It is also possible to generate synthetic sets with a different number of elements from the original data set, but we do not consider that complication here.

Subject to the assumption that the data points are independent in distribution and order (Press *et al.* 1992, Section 15.6), the bootstrap method can answer the question as to the form of the sampling distribution for the statistic, and what is (say) the 68% confidence interval. Replacement is crucial in the procedure, as it guarantees that the synthetic data sets are not simply identical to the genuine data set, i.e. any of the original data points can appear once, not at all, or many times in a synthetic data set. As only the genuine data are used, the bootstrap is an example of a **resampling** method. Superficially, it appears as if we are getting something for nothing from the bootstrap method. This was one reason why it took some time for the bootstrap method to gain acceptance. Nowadays the method is backed by sufficiently rigorous theorems to be regarded as reputable. Since the method involves a random number generator it fits into the class of Monte Carlo methods discussed in Section 9.3.

9.5 Bayesian inference

All of the results presented to date in this book have been within the framework of the **frequentist** approach to statistics. An event's probability is interpreted as the limit of its relative frequency after a large number of trials. An alternative approach to statistics is adopted by **Bayesians**. Bayes' theorem[5] was derived from the general axioms of probability, and relates the conditional and prior probabilities of events A and B:

[5]Named after the eighteenth century English clergyman Thomas Bayes.

$$P(A|B) = P(A)\frac{P(B|A)}{P(B)}, \tag{9.7}$$

where $P(A|B)$ is the **conditional probability** of obtaining A *given* that event B has occurred, $P(B|A)$ is the conditional probability of obtaining B given that event A has occurred, and $P(A)$ and $P(B)$ are the unconditional, or **prior**, probabilities for A and B. The term prior probabilities for A conveys that the probability does not take into account any information about B. Note in particular that A and B do not have to be repeatable events. The vertical bar ('|') in some of the terms in eqn (9.7) denotes 'given'—the information to the right of the bar is taken as being true. In the Bayesian formulation the background information, denoted I, is explicitly included to emphasise that our calculations often make assumptions, and are hence conditional on I.

Inherent in the Bayesian formulation is the evolution of our certainty about a certain hypothesis, H, when more data become available. The prior probability which represents the degree of plausibility of the hypothesis, H, given the

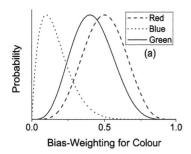

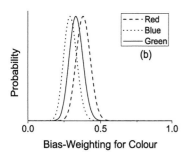

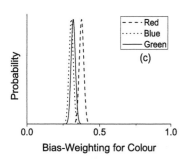

background information, I, is $P(H|I)$. After experimental measurements, some data, D, become available. Using eqn (9.7) we modify the **likelihood** function, $P(H|D, I)$:

$$P(H|D, I) = P(H|I)\frac{P(D|H, I)}{P(D|I)}, \qquad (9.8)$$

which represents the **posterior** probability, i.e. the plausibility of the hypothesis H, given the data and background information.

To illustrate the Bayesian formulation, consider the following computer generated example. A bag contains coloured balls, in the ratio red(40): green(30): blue(30). We consider the evolution of the posterior probability distribution function for the bias weighting of selecting a specific colour as a function of the number of samples. If we denote by G the bias weighting for obtaining a green ball, then $1 - G$ is the bias weighting for not obtaining a green ball. After N samples the posterior probability distribution function, $P(\text{GREEN}|D, I)$, is proportional to[6] $G^a(1 - G)^b$, where a is the number of green balls obtained, b is the number of red or blue (i.e. not green) balls obtained, and $a + b = N$. A similar result is obtained for the bias weighting for red and blue. Figure 9.5 shows the evolution of the (unnormalised) posterior probability distribution functions for the bias weightings for red, green and blue. The number of samples are 10 for (a), 100 for (b) and 1000 for (c). As the number of trials increases the posterior probability distribution functions sharpen and peak at values close to the parent values. The functions have been scaled to have the same maximum value, as it is the shape of the functions which is of interest to us. Note that (i) for a relatively small number of samples there is a broad range of bias weightings for each colour; (ii) as more data become available the centres of the functions evolve much less, and the widths of the distributions become narrower, reflecting our higher confidence in the values of the bias weighting obtained. The evolution of the peak position of the different posterior probability distribution functions shown in Fig. 9.5 as a function of the number of samples is listed in Table 9.2.

An excellent treatment of Bayesian inference in data analysis is to be found in the book of Sivia and Skilling (2006); the book by Cowan (1998) gives examples of Bayesian statistical tests in particle physics experiments; and the article 'Why isn't every physicist a Bayesian?' by Cousins (1995) illustrates the issues involved when a practising experimenter chooses which of the two

Fig. 9.5 The evolution of the posterior probability distribution function for the bias-weighting of selecting a specific colour from a computer generated data set with the ratio red(40): green(30): blue(30) as a function of the number of samples. The number of samples are 10 for (a), 100 for (b) and 1000 for (c). As the number of trials increases the posterior probability distribution functions sharpen and peak at values close to the parent values. The functions have been scaled to have the same maximum value.

[6]This is the binomial distribution function, and the derivation of this result is outlined in Chapter 2 of Sivia and Skilling (2006).

Table 9.2 The peak position of the different posterior probability distribution functions shown in Fig. 9.5 as a function of the number of samples.

Samples	Red	Blue	Green
10	0.5	0.1	0.4
50	0.42	0.16	0.42
100	0.38	0.29	0.33
500	0.378	0.340	0.282
1000	0.376	0.319	0.305
2000	0.3955	0.3095	0.2950

major frameworks to adopt. D'Agostini's article (2003) highlights the role of Occam's razor, discussed in Section 8.7, in the Bayesian approach.

9.6   GUM—Guide to the Expression of Uncertainty in Measurement

We noted in Chapter 1 the potential confusion over the interchangeable use of the terms 'error' and 'uncertainty'. In the 1990s many international bodies which represent professional metrologists, who have responsibility for measurements and standards, published *Guide to the Expression of Uncertainty in Measurement*, or the GUM (1995). Organisations which support GUM include *Bureau International des Poids et Mesures (BIPM)*, *International Union of Pure and Applied Physics (IUPAP)*, *International Union of Pure and Applied Chemistry (IUPAC)*, and the *International Organization for Standardization (ISO)*. The motivation for GUM is to clarify the definition of uncertainty, with the goal of supporting a fully consistent and transferable evaluation of measurement uncertainty. It should be noted that, at present, the use of GUM is not widespread in university laboratories. In this book we have classified errors as either random or systematic, and made no attempt to combine them. In the GUM procedure a *Type A* uncertainty is evaluated by statistical methods, and a *Type B* uncertainty is evaluated by non-statistical methods. The statistical techniques introduced in Chapter 2 for dealing with repeat measurements can be used to evaluate Type A uncertainties; examples of probability distribution functions relevant for Type B uncertainties are the Gaussian and uniform distributions encountered in Chapter 2. After the relevant evaluations, the Type A and B uncertainties may, or may not, be reported separately, in contrast to the combination of the two components, which is always reported. A very readable introduction to GUM can be found in Kirkup and Frenkel's book (2006).

Bibliography

Amsler, C. *et al.* (Particle Data Group) (2008). *Phys. Lett.*, **B667**, 1–1340.

Barford, N. C. (1985). *Experimental Measurements: Precision, Error and Truth*. John Wiley, *New York*.

Bates, D. M. and Watts, D. G. (1988). *Nonlinear Regression Analysis and its Applications*. John Wiley, *New York*.

Bevington, P. R. and Robinson, D. K. (2003). *Data Reduction and Error Analysis*. McGraw-Hill, New York.

Boggs, P. T., Byrd, R. H. and Schnabel, R. B. (1987). A stable and efficient algorithm for nonlinear orthogonal distance regression. *SIAM J. Sci. and Stat. Comput.*, **8**, 1052–1078.

Cousins, R. D. (1995). Why isn't every physicist a Bayesian? *Amer. J. Phys.*, **63**, 398–410.

Cowan, G. (1998). *Statistical Data Analysis*. Oxford University Press, Oxford.

D'Agostini, G. (2003). Bayesian inference in processing experimental data: principles and basic applications. *Rep. Prog. Phys.*, **66**, 1383–1419.

Davison, A. C. and Hinkley, D. V. (1997). *Bootstrap Methods and their Applications*, Cambridge University Press, Cambridge.

Durbin, J. and Watson, G. S. (1950). Testing for serial correlation in least squares regression: I. *Biometrika*, **37**, 409–428.

Efron, B. and Tibshirani, R. (1986). Bootstrap methods for standard errors, confidence intervals, and other measures of statistical accuracy. *Statist. Sci.*, **1**, 54–75.

Feynman, R. P., Leighton, R. B. and Sands, M. (1963). *The Feynman Lectures on Physics*, Volume 1, Addison-Wesley, Reading, MA.

Guide to the Expression of Uncertainty in Measurement (1995). 2nd edition. International Organization for Standardization, Geneva.

Hibbert, D. B. and Gooding, J. J. (2006). *Data Analysis for Chemistry*. Oxford University Press, Oxford.

Higbie, J. (1976). Uncertainty in the measured width of a random distribution. *Amer. J. Phys.*, **44**, 706–707.

Holland, J. H. (1975). *Adaptation in Natural and Artificial Systems* University of Michigan Press, Ann Arbor, MI.

James, F. (2006). *Statistical Methods in Experimental Physics*. 2nd edition World Scientific, Singapore.

Kirkpatrick, S., Gelatt Jr., C. D. and Vecchi, M. P. (1983). Optimization by simulated annealing. *Science*, **220**, 671–680.

Kirkup, L. and Frenkel, B. (2006). *An Introduction to Uncertainty in Measurement*. Cambridge University Press, Cambridge.

Kleppner, D. and Kolenkow R. J. (1978). *An Introduction to Mechanics*. McGraw-Hill, New York.

Levenberg, K. (1944). A method for the solution of certain problems in least squares. *Quart. Appl. Math.*, **2**, 164–168.

Lyons, L. (1991). *Data Analysis for Physical Science Students*. Cambridge University Press, Cambridge.

Macdonald, J. R. and Thompson, W. J. (1992). Least-squares fitting when both variables contain errors: pitfalls and possibilities. *Amer. J. of Phys.*, **60**, 66–73.

Marquardt, D. W. (1963). An algorithm for least-squares estimation of nonlinear parameters. *SIAM J. Appl. Math.*, **11**, 431–441.

Press, W. H., Teukolsky, S. A., Vetterling, W.T. and Flannery, B.P. (1992). *Numerical Recipes in Fortran 77*, 2nd edition. Cambridge University Press, Cambridge.

Reed, B. C. (1989). Linear least-squares fits with errors in both coordinates. *Amer. J. Phys.*, **57**, 642–646, Erratum *Amer. J. Phys.*, **58**, 189.

Reed, B. C. (1992). Linear least-squares fits with errors in both coordinates. II: Comments on parameter variances. *Amer. J. Phys.*, **60**, 59–62.

Rees, D. G. (1987). *Foundations of Statistics*. Chapman and Hall, London.

Silverman, M. P., Strange, W. and Lipscombe, T. C. (2004). The distribution of composite measurements: How to be certain of the uncertainties in what we measure. *Amer. J. Phys.*, **72**, 1068–1081.

Sivia, D. S. and Skilling, J. (2006). *Data Analysis a Bayesian Tutorial*, 2nd edition. Oxford University Press, Oxford.

Squires, G. L. (2001). *Practical Physics*. Cambridge University Press, Cambridge.

Taylor, J. R. (1997). *An Introduction to Error Analysis*. University Science Books, California.

Winer, B. J. (1962). *Statistical Principles in Experimental Design*. McGraw-Hill, New York.

Index